4 April 2006

To Helen and Elton

I hope you enjoyed the reading of these book, that among other things brings the use of tobacco as part of the religious rituals of pre Columbian cultures..

[illegible]

Between Puffs

A History of Tobacco
Two Thousand Years of Tobacco Use

Nicotiana
Tabacum
Ruben Oropeza

Between Puffs

A History of Tobacco

Two Thousand Years of Tobacco Use

by

Ruben Oropeza, M.D., F.A.C.S.

RIVERCROSS PUBLISHING, INC.
Orlando

 For information address RIVERCROSS PUBLISHING, INC., 6214 Wynfield Court, Orlando, FL 32819, or editor@rivercross.com.

ISBN: 1-58141-092-1

Library of Congress Cataloging-in-Publication Data

Oropeza, Ruben, 1930–
Between puffs : a history of tobacco : two thousand years of tobacco use / by Ruben Oropeza.
p. cm.
Includes bibliographical references and index.
ISBN 1-58141-092-1 (alk. paper)
1. Tobacco habit—History. 2. Tobacco—History I. Title.

HV5730.076 2005
394.1'4—dc22

2005046441

1 2 3 4 5 6 7 8 9 10 10 09 08 07 06 05

Dedicated to the memory of my mentor, teacher and friend, Dr. George T. Pack, surgeon par excellence.

To my wife Joyce and my children David, Marcelle, and Victoria, for their support, patience and understanding throughout my medical career.

To the memory of my parents with gratitude, love and respect.

To my long-time friend, Dr. Nemethala Ghossein

ACKNOWLEDGEMENTS

The author wishes to acknowledge his sincere gratitude and appreciation to Ms. Mary Miller for her constructive critique of Part I of this book.

Ms. Miller is a Vincent Scully Professor of the History of Art and Master of Saybrook College, Yale University.
She is the author of the Art of Mesoamerica, Maya Art and Architecture and many scholarly works on Maya culture. In her most recent book Courtly Art of the Ancient Maya (2004), she illustrates the practice of tobacco use by these people through artifacts found in present-day México, Belize and Guatemala.

The author also wishes to express his appreciation to Gwenn Lewis Norman for her enthusiastic assistance in obtaining books and other literature related to tobacco.

My appreciation to Roger Hagan for reviewing the manuscript.

My thanks to Marjorie J. Sneed for her wholehearted and tireless dedication in the typing of this manuscript.

To Steven Hoare and the other editors at Rivercross Publishing under Josh Furman who were most helpful and patient in making this book possible.

'Columbus came looking for the Great Khan, silver, spices and gold and found tobacco instead'

Nebur Azeporo (1930–)

'Farewell to tobacco'
'Filth of the mouth and fog of the mind
Hemlock, aconite—(tobacco)
        Nay, rather
Plant divine of rarest virtue'

Charles Lamb 1909 London

CONTENTS

PART II

FOREWORD

This remarkable book documents, in a readable fashion, the history of tobacco use worldwide over the past two thousand years. In particular, part two of the book concentrates on the last 500 years of this history since the ''discovery'' of the Americas by Christopher Columbus in 1492. This part of the history includes the introduction of tobacco from the New World (Hispaniola) to Europe and other parts of the world leading up to the present time when this lethal commodity is known to be the cause of more than 5 million deaths each year worldwide and is the number one cause of preventable deaths in America where it kills 440,000 people each year.

Perhaps the most riveting and least known part of this history of tobacco as told in this book is that beyond the devastation and death caused by tobacco, the author reveals that the production, promotion and sale of tobacco has been a significant determinant of human events and world history over the past 500 years. This history has played itself out form the time of the discovery of America, which led to the colonization of North and South America by Europeans. The slave trade, for example, was driven by the need to have a workforce in America to provide hands that would grow and pick tobacco, a commodity which became exceedingly high in value filling the coffers of a number of European countries—Spain, Portugal and England in particular. Furthermore, the author provides evidence to suggest that the American Revolution itself was in part driven by a rebellion over tobacco by colonists who sought reasonable compensation for the commodity which they produced.

The development of the cigarette in 1885 stimulated the mass production of the product and the advent of the large tobacco corporations which took their place among the big businesses of the Americas and elsewhere.

Paradoxically, related to this history, ruling authorities and governments over several centuries and to the present day have also become addicted to tobacco.

The author underscores the challenges faced by addicted smokers, half of whom will die due to tobacco, but also by "addicted" governments whose economies are far too dependent on taxes generated by tobacco even though the weed kills five million people worldwide each year.

Harold P. Freeman, MD, FACS
Professor of Surgery
Columbia University
Past President American Cancer Society

A CHRONOLOGY OF THE FIRST OBSERVATIONS OF TOBACCO USE BY EUROPEAN EXPLORERS.

The West Indies

1492 Christopher Columbus—The Taíno, Arawak and the Carib.

Mesoamerica

1517 Hernando de Córdoba
1518 Juan de Grijalva—The Maya and Totonac
1519 Hernán Cortés—The Aztec
1539 Viceroy Juan de Mendoza—The Tarasca

South America

1499 Amerigo Vespucci—Amazon, Orinoco and Margarita Island
1500 Pedro Álvarez Cabral—Tupinamba (Brazil)
1531 Francisco Pizarro—Inca and the East Coast of South America
1565 Girolamo Benzoni—North Coast of South America and Central America
1555 André Thevet—Brazil
1595 Sir Walter Raleigh—Orinoco—(Guiana)

North America

East Coast

1524 Giovanni Verrazano—Manhattan (New York) and Narragansett (New England)
1534 Jacques Cartier—Hurons, St. Lawrence River (Canada)
1562 Jean Ribaut—Timucua (South Carolina)
1563 René Laudonnièr and Lemoyne—Timucua
1562 John Hawkins—Carolina-Virginian people

1585 Thomas Hariot—Carolina-Virginia
1573 Sir Francis Drake—Carolina-Virginia

The Great Lakes and Mississippi River

1603 Samuel Champlain—Huron, St. Lawrence River (Canada)
1666 Jacques Marquette—Ottawa, Winnebago (Illinois)
1679 Louis Hennepin and René Robert Cavelier—Sioux, Upper Mississippi River

Southwest

1528 Alvar Núñez Cabeza de Vaca, Florida and Arid-South tribes (present-day Texas-New Mexico) Sonora-Chihuahua Desert

The West Coast

1578 Sir Francis Drake—Cultures of the Coast of California
1728 Vitus Jonassan Bering—Kayak (Alaska)
1785 Count Francois de la Perousse—Tlingit (Alaska)
1816 Captain Corcoran Katzebue Expedition (Alaska)

Northwest

1804 William Clark and Meriwether Lewis—The Missouri Expedition
1810 Alexander Ross—Nez Perce (The Rockies)
1864 Francis Parkman—Pawnee (The Oregon Trail)

The Great Plains

1830 George Catlin—The Sioux, Mandan, and others

NAMES OF TOBACCO

The tobacco plant has been known by many different names by different cultures throughout history. The names given to tobacco by pre-Columbian cultures of the Americas include:

Kahiha (Caribe, peoples of the Caribbean)
Upponuc (Algonquin, peoples of Virginia 1584)
Yetl & Picetl (Aztec)
Sagri (Inca)
Kiyekta (Huron)
Cohiba (Caribe)
Yuri (Columbian people)
Ureit (Trinidad people)
Yetl (Mexica people)
Cojala (Taíno)
Petum (Brazil)

The first name given to tobacco by the Spaniards in the 15th century was *Herba Sancta* or *Hierba Santa* (sacred weed).

In the 16th—17th centuries tobacco was varyingly referred to as:

Angoulmoisine (A. Thevet 1558). After his place of birth
The magic herb
The noble weed
The new mischief
The healing herb
The sovereign herb
The mighty leaf
The poisonous weed
L'Herbe du Grand Prieur
Herbe Medicée (after Catherine de Medici)
Herba Nicotiana
Herba Di Santa Croce
Herba Tornabuoni
The filthy weed
The stink (King James I)
Nicotiana Inferta (the weed of hell)

The accursed weed (Florence, Italy)
The enchanted herb (Napoleon Bonaparte)
Herba Panacea (Dutch)
Sana Sancta Indorum (Gerard's Herbal of 1597)
Holy Herb Nicotiana (William Lily, Queen Elizabeth's court poet)
Divine tobacco (Edmund Spenser, a poet friend of Sir Walter Raleigh)
Angoulême (Thevet, France, 1558)
Panacee antartique
Herbe de L'Ambassadeur (France)
Herbe Sainte
Herbe Sacrée

or simply, the weed.

INTRODUCTION
THE MAGICAL WEED

Tobacco was cultivated throughout the entire New World before Columbus' arrival. How it came to be cultivated is a mystery, but we may speculate that there was something special about tobacco, some discovery that Amerindians made about its nature or the ease of its cultivation that led to its extensive cultivation. Tobacco grew in virtually every type of soil and climate, and required little growing space. Like all hallucinogens it was revered, but unlike many its effects were relatively mild, which must have added to its allure.

Both sacred and aromatic, tobacco probably was first used as a form of incense during religious ceremonies. At some unknown later period, Amerindians began to employ tobacco medically, as a curative for many ailments. Gradually, over time, they started consuming it during political and civic rites, and on both formal and informal social occasions. These uses were interrelated, emanating from the belief that tobacco was a sacred herb. Within this context each tribe, nation or civilization had its own uses for tobacco, its own preferred methods of consumption, its own additives to enhance the taste or smell, to speed up or otherwise modify the effects. The Indians devised three ways of curing, or drying, the plant's leaves—by fire, drying in the sun or air drying. In the end, however, whatever the mode of consumption or production, perhaps the most astonishing fact about tobacco was its deep penetration into all aspects of Amerindian cultures.

Tobacco has had a very strong impact on the lives of people and has penetrated into many aspects of society. In the 19th century, Napoleon carried a snuffbox to the battle at Waterloo. In the 20th century, Stalin smoked a pipe, Churchill a cigar, Roosevelt and Mao cigarettes. Among World War II leaders, only Hitler was a nonsmoker. As we begin the 21st century, tobacco is a huge social and health problem.

Efforts are underway to fight the problem. Television ads exhort us to quit smoking by wearing nicotine patches. Recent films like The Insider expose the collusion between tobacco companies and the media to prevent the truth from reaching the public about the dangers to health caused by smoking. States have pushed through class-action suits against tobacco companies, calling for them to bear the medical costs of treating people with cigarette-induced cancer. Teen smoking, Third-World smoking, the increase in smoking by women, and the costs to society in terms of medical expenses and the loss of productivity in the workplace are matters of grave societal concern.

Scientific research has confirmed the life-threatening aspects of tobacco consumption while also identifying nicotine as a highly addictive substance. Writings on the subject increasingly focus on the dangers of tobacco to smokers and nonsmokers (via secondhand smoke) alike, as well as to the environment.

Bombarded with these facts we may think that tobacco use has always troubled society but, in fact, tobacco has not always been viewed negatively. From the moment when Christopher Columbus and his crews became the first Europeans to encounter this native plant of the Americas and brought it back with them to Europe, tobacco has been viewed with deep ambivalence. Considered by some as an instrument of the devil, tobacco has far more frequently been revered as "the sacred weed," a miracle medicine capable of saving, not condemning lives. Whatever one's view, tobacco has profoundly influenced world culture and illuminated issues of class, gender, law, status, economics, business, religion, power and privilege within the world's societies.

For the past five centuries unsuccessful attempts have been made to stem the use of tobacco. Christian and Muslim leaders proclaimed in the 16th and 17th centuries that tobacco use was contrary to the teachings of the Bible and the Koran. Secular authorities also condemned tobacco use as being immoral or simply disgusting. King James I of England called tobacco a "filthy weed" in 1603. Louis XIV of France disliked all tobacco use, while Napoleon Bonaparte disliked some aspects of tobacco use but not others. It wasn't until the 20th and 21st centuries, however, that tobacco use was campaigned against on the basis of being detrimental to health.

The hope to discover why it has thus far proven impossible to curb tobacco use in the face of indisputable evidence that tobacco is a serious health hazard led to this author's historical review of the use of tobacco by different cultures for the past twenty centuries.

PART I

1.

THE MIGHTY LEAF

Tobacco in Pre-Columbian Cultures

The first name given to tobacco by the Spaniards in the 15th and 16th centuries was *Herba Sancta-Hierba Santa* (sacred weed), (Fig. 1) but many other names have been given to this plant.

Tobacco is the common name given to the plant *nicotiana*, but tobacco (or *tobago*) was the name given by the Taínos (people of the Caribbean) to the act of smoking the dried, rolled leaves of this plant, rather than to the plant itself. The Spaniards later called this plant the tobacco plant.[1] The word "tobacco" is of Guaraní or Taino origin (West Indies)[2] and "tobacco" was the word commonly used by Oviedo in his publications in 1535.[3] Bartolomé de Las Casas in 1552 used the word "tobacco" in reference to the first cigars seen in Cuba, as mentioned in the Oxford English Dictionary: "the dry leaves which are rolled that are kindled at the end and used like a rude cigar".[4] Las Casas, Sahagún, Hernández, Oviedo and others mention two different types of tobacco in their 16th-century descriptions: *yetl* and *picietl*. The Tarasca knew it as *sinchakua* (n.tabacum) and *anda mukua* (n.rustica).

Juan de Cárdenas, 16th-century physician and administrator of the New Indies in Hispaniola, made one of the first detailed descriptions of the tobacco plant after the discovery of the Americas. "This precious herb," he wrote, "is so general a human need not only for the sick but for the healthy. Some are accustomed to taking it in small clay or silver pipes or those of hard wood, others wrap the tobacco in a corn husk or in paper or in a tube or cane." Cárdenas discussed the "medicinal qualities" of tobacco more than its botanical characteristics, and influenced

Fig. 1. Herba sancta (sacred weed). A 16th c depiction of Nicotiana Tabacum by an unknown author.

Monardes and others about the qualities of the plant.[5] He prescribed tobacco for many illnesses while in Santo Domingo, stating:

> When I start thinking who invented the sucking of tobacco smoke [*Piciette-Picente*] and since there is no mention of who might have said it, I suspect that an angel advised it to the Indians or a devil. . . . That it was an angel

makes sense because it frees us of so many sicknesses, that really it seems to be medicine of the angels, and that it also seems to be by the devil when we start looking at those that are smoking it [*Chupandolo*] that throws it [smoke] to the mouth and through the nose, a stinky smoke, that he looks like a volcano, or the mouth of the 'inferno' but whoever might have invented [discovered] it seem to me that it should be properly used for our diseases, medicine of heaven, in spite of being harmful, pernicious and stinking if not properly used.

André Thevet, a Huguenot Calvinist who spent three months in Brazil from November 1555 to January 1556, described the use of tobacco, calling it *peton* (from the Tupi-Guarain word *pituma* or *pitima*). He witnessed a native ". . . who rolled the leaf and wrapped it in a large palm leaf to

Fig. 2. 'The Christians There Today' (Brazil). Natives of Brazil smoking a cigar similiar to Columbus' first description in 1492. André Thevet 1557. New York Public Library (NYPL)

the length of a candle." (Fig. 2) Thevet tried tobacco himself, with bad effects. He described pipe smoking in Canada in 1575, and brought the secret of tobacco to his native town of Angoulême in France. Thevet refers to Nicot as *un quidam* (someone) who never traveled abroad and who falsely claimed to have been the first to bring tobacco to France.[6]

Oviedo mentions, "This herb is about four or five palms in height with broad, thick, soft and velvety leaves. They have it in great prize and they planted it for its use."[7]

México and Central America, according to the Russian botanist Vavilov,[8] were the origin of cultivated medicinal and other plants, including tobacco, in the New World.

Early botanists identified two different tobacco plants: *nicotiana tabacum*[9]; and *nicotiana mexicana*, called *yetl* in the Nahuatl language of the Aztec[10] and *sinchakua* in Tarasca[11] (both were cultures of present-day México). *Nicotiana rustica* also was present in México at the start of the Spanish colony in the 16th century. Nicotiana rustica was called *pisietl*, which means "small."[12] The leaves of this plant are smaller than those of the nicotiana tabacum or nicotiana mexicana. In the Maya culture tobacco was called *kutz*.

The people of North America—Mandan, Hidatsa, Arikara, and Dakota —smoked *nicotiana bigelovi*, and N. quadrivalis which was as strong as nicotiana rustica. The people of North America's southwest and the Sonora Indians of northwest México smoked and chewed *nicotiana attenuata* and *n. obtusifolla*.[13] The Indians of the North America's northeast grew mainly nicotiana rustica until 1610 when John Rolfe (husband of Pocahontas) introduced nicotiana tabacum to Virginia. Rolfe probably brought the seeds from Trinidad.

The colors of the flowers of the tobacco plant vary from white to red. (Fig. C1) Some of these plants are used in gardens for ornamental purposes because of the beauty of their trumpet-like flowers. (Fig. C2) In the 16th century nicotiana was a popular ornamental plant in the gardens of Portugal and Spain. In Iran they cultivate n.tumbak (n.alata), which is used also in Turkey where it is smoked in narghiles. N.sylvestris and n.alata grandiflora are ornamental plants of the Solanacea family. (Fig. C3)

Nicotiana tabacum, together with seventy-two other genera of tobacco plants, is a part of the *solancea* family. Many of the plants that belong to this family contain an alkaloid with narcotic stimulant or hallucinogenic properties. Tobacco was often mixed with the seeds of other plants such as morning glory (*Ipomea violacea*), ololihuqui (*Ipomea corynbosa*). Both hullucionogenic. (psychodelic)

Fray Bernardino de Sahagún in the 16th century made the distinction between nicotiana rustica (or picietl) and nicotiana tabacum (or yetl). He wrote of nicotiana rustica that when "placed in the mouth it produces dizziness and stupefies." Regarding nicotiana tabacum, it "is sweet as a kind of ceremonial incense."[14]

2.

TOBACCO AND SHAMANISM IN AMERICA

Origins and Legends of Tobacco

We can trace the origin of tobacco smoking for at least 1,500 years through the artifacts, ceramics, pipes and murals of pre-Columbian American cultures such as the Olmec-Totonac, Maya, Toltec, Aztec, Mixtec, Tarascan and others. Further information has been obtained from the Maya Codices, the polychrome plate of Uaxactum (Guatemala), the Stele in the Temple of the Cross in Palenque Chiapas, the mural of Tepantitla (Teotihuacan, México), through the extensive documentation of Mayan artifacts described in the books by Frances Robicsek, and more recently by Mary Miller, (Courtly Art of the Ancient Maya) and from the recent find by the author at the museum Rufino Tamayo in Oaxaca of two Totonac figures—one smoking a cigar and another chewing tobacco.

We know through legends propagated by indigenous cultures that the use of tobacco was important in their daily lives. Tobacco even figures in a Taino (Caribbean) legend of the creation:

A son of Itaba-Taunana by the name of Caracaracoel asked for casaba (from a root of the yucca plant) from his grandfather Ayanamaco or Basa Manaco. The grandfather, however, was bothered by the request and hit the child on the back with a bag containing the powder of cojala (Taíno name for tobacco) causing a painful swelling on the child's back. The brother opened the swelling using a manaya (hatch) and a female turtle came out and built its home on her top, thus creating the earth. This legend gives importance to tobacco in the creation of earth and man.[1]

The story goes that when the Spaniards first observed the Taíno (the people of the Island of Hispaniola, now Haiti and the Dominican

Republic) smoking tobacco, the Spaniards attributed dual purposes to tobacco. "One was attributed to the angels, and the other to the devil. The one to the angels was the power to cure many illnesses, and the other was given to the devil because when smoke came out of their mouths and noses they resembled volcanoes in eruption."[2]

A legend of the Susquehanna Indians of North America describes the creation of tobacco:

> Once upon a time, long, long ago, when the Red Indians of America ate only meat, a group of hunters paused during a shooting party, to partake of a meal. A deer they had just killed had been roasted over a fire and they had eaten to their hearts' content. In deep silence, the grotesque nature of the wild mountainous country lay around them. Suddenly, on the slopes of one of the neighboring mountains, they saw the outline of the body of a beautiful young woman, slowly moving toward them. It was a wood nymph of awe-inspiring beauty. They were all terrified, and supposed the nymph to be hungry and allured by the smell of the venison. They took some tender haunches of venison to a neighboring plateau, offering the tongue of the deer. They were right; the apparition partook of it, and proved to be pleased by the remarkably good flavor. When she was satisfied, she spoke to the Indians. "Your kindness will be rewarded, come back to this spot after the moon shall have risen 13 times. Then you shall find your reward." And so it happened, when they came back after 13 moons, they saw that on the spot where the apparition's right hand touched the ground maize was growing, and where her left hand had touched the ground, beans were growing, and where she had sat, tobacco was growing. The plants were tended and finally the seeds were gathered, as were the leaves of the tobacco plant, which seemed to be sweet smelling and tender. The following years the seeds were sown again, and thus these plants spread all over the earth.

And so like many other legends, the truth of the origin of tobacco is uncertain.[3]

Possible Explanation of the Religious, Ritual and Medicinal Use of Tobacco by the Indigenous People of America

Primitive cultures, as we know from Herodotus, found supernatural reasons for unexplained and intriguing phenomena. They thought that supernatural spiritual forces were at play behind such phenomena as fire, wind, and smoke. When the Aztecs of México observed the eruption of the Popocatépetl volcano (the "smoking mountain"), followed by fire and smoke, they interpreted the event as proof that the gods were angry with them for their bad deeds and determined that sacrifices had to be made to appease the gods. Smoke had divine connotations for the Aztec because it drifted to heaven to the gods. The smoke from tobacco, by analogy, might have had the same spiritual meaning. (Morley)

The attributes of the tobacco leaf must have been discovered by some of these cultures through trial and error. They chewed the leaves to taste the flavor and test its nutritional qualities, or to test its stimulative or mood-altering effects. When they found that the rolled leaves of this plant produced not only fire, scent and smoke, but produced a mood change on those who used it, tobacco became part of their religious rituals and was given sacred attributes.[4] From then on the smoking of tobacco was used by priests during religious ceremonies. According to their beliefs, sickness was caused by possession by evil spirits. Tobacco, being considered a sacred plant, was used during these ceremonies by all American cultures to conjure up the evil spirits, and thus was given medical attributes.

Their tobacco contained up to ten times the amount of nicotine as our present tobacco. The stimulant (psychotropic) effect of such strong nicotine caused the priests to hallucinate while they were performing their religious ceremonies. This was of great importance because they believed that they were transported to the heavens and could be possessed by the gods in this state.[5] The rest of the population then took up this habit, which led to addiction not for religious purpose, but for pure personal enjoyment.

The people of Mesoamérica (México, Guatemala, Belize and Honduras) smoked tobacco in pipes[6] or in the rolled leaves of tobacco (cigars).[7] In the Totonac-Maya culture they also chewed tobacco. Most of the pipes in Mesoamerica are carbon-dated from the 8th century to the 14th century A.D.[8] Indians of North America smoked tobacco in pipes. Some people of South America smoked it in pipes (The Chaco) or they chewed it (as in Margarita Island, Venezuela). There is a logical explanation for these different methods of using tobacco. The aborigines of North America smoked nicotiana rustica. The leaves of this plant are very small and unsuitable to roll into a cigar. On the contrary the leaves of nicotiana tabacum are large and thus suitable to roll into cigars. Furthermore, tobacco in some of the cultures of North America was sparse and could not be wasted, while the cultures of Mesoamerica and South America (such as in the Orinoco) had abundant tobacco crops. Pipes are much more sparing and less wasteful of tobacco than cigars.

These cultures also used enema syringes for the use of tobacco, coca and hallucinogenic drugs.[9] (Fig. C4) Evidence of snuffing in pre-Hispanic cultures has been documented.[10]

Shamanism in America

Shamanism existed in Asia before man first set foot on the American continent. It has been postulated[11] that Shamanism came to America as part of the cultures that migrated to the Americas from Asia. They knew hemp, which was consumed by throwing the seeds or leaves of the plant in fires and inhaling the smoke. That custom also was practiced by cultures of the Middle East and India. This archaic shamanistic practice united all aboriginal Native American cultures. Tobacco was incorporated by Asian Shamans in the 17th c.

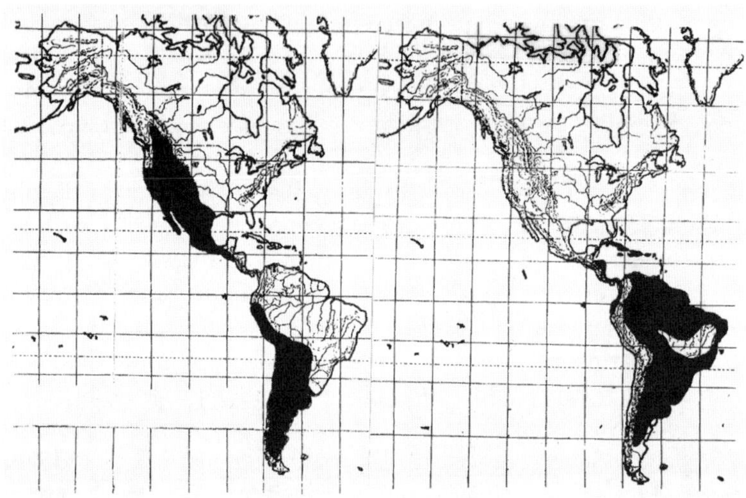

Fig. 3. Geographic distribution of wild tobacco. Darkened areas show where wild tobaccos grow; about sixty species appear in the two mountainous areas of America, the heaviest concentration being in Chile and Argentina. Between Ecuador and Guatemala, natural distribution is interrupted, and there are no strictly wild tobaccos in nothern South America and the West Indies; species diminish in numbers northward through British Columbia.

At right: Darkened areas in Central and South America show Indian cultivation of *Nicotiana Tabacum,* the superior species. Spread by man as part of the Amazonian agricultural complex of maize, manioc, peanuts, etc., in migrations of the Arawak and Tupi-Guarani tribes, it is today's tobacco of commerce. It originated as hybrid on the eastern slopes of the Andes. NYPL

Tobacco is native to America. It grew wild in all climates, from the Arctic climes of the north to the tropics of Meso-South America; from the high altitudes of present-day México City and the Andes (8,000 feet above sea level) to the jungles of Yucatán, México, the Orinoco-Amazon basin, and the Caribbean islands. (Fig. 3–4) What better, safer and milder psychotropic drug to use than tobacco? It was sometimes mixed with whatever hallucinogens were available in a particular area—*peyote, ololiuhqui,*

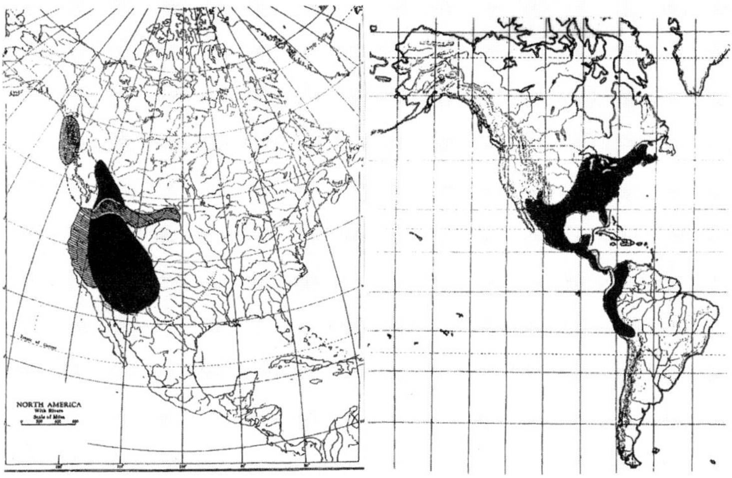

Fig. 4. Geographic distribution of other varities of wild tobacco. Map of North America, showing wild tobaccos in the Northwestern Frontier. *Nicotiana attenuta* (black) reaches farthest north of the indigenous tobaccos; *Nicotiana bigelovii* (cross-hatching), a wild and semi-domesticated species in California, comes next. Also shown in cross-hatching is *N. quadrivalvis* which survives in North Dakota among the Mandan and Hidasta Indians.

At right: Map of Western Hemisphere showing, in black area, the extent of the spread of *Nicotina rustica* which was aborginally cultivated in México and Central America. In Virginia, it was soon replaced in English cultivation with *Nicotina Tabacum* introduced from the Orinoco. NYPL

mushrooms, *a.peregrina*, henbane, and many other plants that were well known to these cultures. They experimented by chewing innumerable plants or their seeds to determine which had these properties.

The power of the shaman during his religious and/or medicinal ritualistic ceremonies was related to the power of these plants to change his vision or perception of things that only he could see or hear. Then, after he had taken this trip to the underworld or to heaven and communicated with the "Great Spirits" that controlled the supernatural forces of this world, he would be possessed with magical powers of divination, healing or any other power that he might deem necessary for a particular religious or civil ceremony. (Fig. 5) With time not only tobacco, but other artifacts such as pipes or urns used during these ceremonies, became sacred. These include the Calumet Pipe of the Sioux or artifacts to carry the tobacco—the calabash[12] for the Aztec and other cultures, or the pouch or bundle that was hung on the necks of the Maya shaman of Yucatán Totonac, Huichol, Pawnee and other American Indians. (Fig. C5)

The shaman would go through an intensive purification period, especially when initiated. The Taino and people of the Orinoco-Amazon basin

Fig. 5. Curing the Sick. First depiction of a healing ceremony in Brazil. Shaman blowing tobacco smoke in the Sick's Face. A. Thevet 1555. Brazil. N.Y. Public Library (NYPL)

would fast for about two weeks, taking only tobacco juice, probably mixed with hallucinogens. They would go into a state of trance and stupor where they would see flying birds, jaguars or serpents. Under the influence of hallucinogens, they also would see psychedelic colors with different patterns. They would hear sounds or strange voices as well. They were forbidden to have sexual intercourse during this time of purification.[13] After they were "cleaned" or purified, they would insert a spatula in their throats before the ceremony so they could vomit and have an empty, "clean" stomach.

During the rituals they smoked tobacco in large cigars or they smoked tobacco in pipes (Aztec, Tarasca, Sioux, Iroquois and other people of North America). They would blow the smoke to the four cardinal points, or only to three (as did some of the Orinoco people). (Fig. 6)

Shamanism is still practiced by many Indian cultures of America.[14]

Fig. 6. Wishiratu shaman of the Orinoco blowing his long tubular pipe to the Kanobo spirits who inhabit the mountains at the end of the world and had to be satisfied with tobacco smoke. Johannes Wilbert. From *Flesh of the Gods* by Peter Furst.

3.

THE EUROPEANS' DISCOVERY OF TOBACCO

"Here Bjarna Grimolfson first drank cloud-Brother through child-of-land-and-water." (An inscription carved in the runic letters of Scandinavia on rocky ledges in Vinland, A.D. 1250.) This Viking inhaled his first tobacco smoke through a reed stem in the Wilkbur Pasture.

The Icelandic Sagas 1250

Christopher Columbus

Christopher Columbus (1451–1506), one of the world's great navigators, sailed with three ships from Palos, Spain, on August 3, 1492, and landed in San Salvador (Bahamas) on October 12, 1492. He founded the first European settlement in the Americas on the north coast of the Island of Hispaniola (Dominican Republic-Haiti).[15]

The great design of reaching the orient through the western ocean had been in the minds of many navigators. Much earlier Biarni Henulfson and Leif Ericson (1000–1010) the Vikings had reached the coast of Newfoundland. The Vikings were probably the first Europeans to smoke tobacco in America.[16]

Exploration of the west coast of Africa and the islands of the Atlantic—the Azores and Canary Islands—had started in 1430 when Prince Henry the Navigator built the Great Portuguese Fleet for the exploration of the coasts of West Africa. The Pope in 1455 had given Portugal the sole right of exploration of this coast.[17] Some of the greatest sailors of Europe came from Genoa, Italy. These seamen had been engaged in looking for new markets for their products and importing goods from Africa and the Mediterranean.

One of these navigators was Christopher Columbus, who was born near Genoa and had extensive experience as a sailor. Columbus was a renaissance man with a sense of adventure and strong faith in himself. His mind was as free and large as the ocean that he was going to conquer. He was not deterred by the tales that this ocean had monsters and that there was an abyss—known as the Mar Tenebroso (dark and gloomy) Incognito (unknown)—from which he could not return.[18] (Fig. 7)

A catalyst for Columbus's first exploration of the western ocean was the conquest of Constantinople (Istanbul) by the Turks in 1453.[19] This had made European commerce with the Middle East, India and beyond extremely difficult. The Turks controlled commerce with the East and imposed heavy taxes on those who wanted to cross their territories. There no longer was free access to the Black Sea (Genoa Lake) and the lands of India. Genoa, Venice, Portugal and Spain had a vital stake in the Mediterranean trade routes after losing this avenue of navigation to the East. New routes had to be found to bypass the Turkish sultans.

Columbus was not the only navigator looking for new routes and new lands. Portugal had one of the best navigators of that time, Pedro Álvares Cabral, who eventually found his way to Brazil (1500) soon after the discovery of the West Indies by Columbus in 1492 and the discovery of North America by another navigator from Genoa, John Cabot, in 1497.

When Columbus returned from an exploration to Guinea in Africa, he had made up his mind to explore the uncharted sea looking for the Spice Islands, the land of Cathay (China), and Cipangu (Japan).[20] He had learned from the Portuguese that long voyages of exploration on the coast of Africa were possible, and he was confident that with some of his existing ships and with God's assistance he could find a western route to the East as well. To finance his voyage to the Indies he had to convince the king and queen of Spain that the enterprise was sound and feasible, and that the vast riches described by Marco Polo centuries before would be found.

Columbus broguht with him the Paolo Toscanelli letter when he was granted the royal audience. This letter argued that the earth was round and that the lands of the East were at a shorter distance to the west than the Europeans had thought up to that time (5,000 miles). The route entailed going through the Island of Antilia (a nonexistent island).[21] Columbus himself calculated the distance to be only 2,500 miles. Fortunately the continents of America happened to be in his way, otherwise he would have never succeeded in his enterprise.

Columbus pleaded in vain for ten years[22] to be financed for this trip by the Courts of Spain while staying on and off in La Rábida Monastery (Fig. C6) until he gained an ally in Luis de Santangel, keeper of King Ferdinand's private purse. The king finally consented and Columbus was issued a passport in Latin stating that he was in command of three caravels (Spanish-Portuguese ships) to explore the Indies. He was given three letters of introduction—one for the Great Khan, Emperor of Cathay, and two blank letters to be filled in as needed. Columbus was prepared for the enterprise and was named Admiral of the Ocean Sea with three caravels under his command—the *Santa María*, (Fig. C7) *La Pinta* and *La Niña*. The armada left from Palos, Spain, on August 2, 1492. Columbus went onboard the *Santa María* on Friday, August 3, and at the break of day made the signal to get underway. The voyage would take two and a half months before they landed in (Guaraní now San Salvador, Bahamas) on October 12, 1492, after making only one stop in the Canary Islands for provisions and repairs to one of the ships.

Most of the information regarding the voyage to the New World—or the Indies, as Columbus named them—comes from Columbus's son, Ferdinand (1535), and Fray Bartolomé de Las Casas[23] who went to the New World with Columbus on his second trip in 1493 under the command of Nicolás de Ovando (Las Casas's uncle had sailed with Columbus on the first trip). Las Casas wrote *Historia de las Indias* (1515), long after Pané and others had given their accounts of these trips. Peter Martyr also gave an account of this voyage (*Orbe Novo*). Fernández de Oviedo y Valdez, an intimate friend of the Infante Don Juan, son of Ferdinand, went to America in 1513 as a controller of the gold diggings of Darién (near present-day Panama), stayed for 34 years, and wrote a history of the New World.

Christopher Columbus himself kept daily logs of his voyage[24] and gave an account in a journal entitled *The Journal of His First Voyage*. Tobacco was mentioned for the first time in two letters sent to King Ferdinand on March 1, 1493, while Columbus was in Lisbon upon his return from his first voyage to America.

Voyage of Discovery

Landfall on October 12, 1492

"In the name of God make Sail."
(The queen's blessing to Columbus)[25]

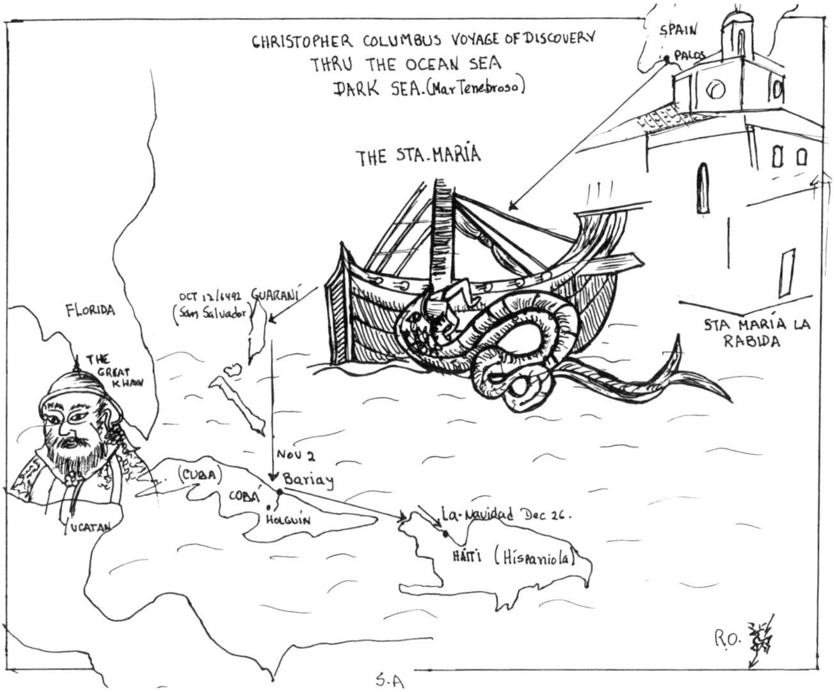

Fig. 7. Columbus' voyage of discovery. A rendering by the author.

Soon after his landfall on October 12th, Columbus became acquainted with tobacco when the dry leaves were offered to him by the natives. He could not identify the leaves at that time, but he was able to identify them later, in a second encounter. While navigating close to the Island of Rum Cay in the Bahamas, Columbus picked up a native in a single canoe who was bringing along glass beads, two small coins (obviously the native was coming from San Salvador where he had met Spaniards), a lump of a bright red-colored substance (*tierra bermeja*) for use in coloring shells or painting the skin, and "some dry leaves which must be something much valued by them, since they offered me some in San Salvador as a gift."[26]

The first account of tobacco written in a European language is probably that of Columbus on October 15, 1492, when he made a note in his journal after reaching the beaches of San Salvador. "The natives brought fruit, wooden spears and certain dried leaves [tobacco] which gave both a distinct fragrance," wrote Columbus. According to Navarrete, "The Spaniards ate the fruit but threw the leaves away."[27]

On October 28th Columbus landed at Bahía Bariay in the beautiful Orient Province of Cuba that Columbus had named "Juana" after the daughter of Queen Isabella. The island was called Cobana or Colba by the Taino people.[28] Columbus found a village there and expected to obtain information on where gold could be found and how to locate the Grand Khan. (The Spaniards had captured Lucayan natives while exploring the first islands, and one of the men became the Spaniards' interpreter on this first voyage. On the next voyage, in 1493, the Spaniards' baptized their interpreter as Diego Colon.) The Spaniards began a brisk trade with the natives, and then what followed was a series of misinterpretations. Columbus inquired about El Gran Can (The Great Khan),[29] but to the natives it sounded like *Cubanacan*, which in Taino means Mid-Cuba. Columbus showed the natives the gold pieces that he had brought along and asked where to find it. The natives, thinking that Columbus was looking for their *cacique* who resided in the inner part of Cuba (Holguín) about twenty-five miles inland from the port of Gibara, suggested through sign language that the Spaniards should come along to meet their chieftain.[30]

Columbus had brought along with him on his caravel an Arabic scholar, an intellectual "Idler" by the name of Luis de Torres. Torres was a Jew who had converted to Christianity (then called a *converso* or *marrano*).[31] He spoke Arabic, Hebrew, Chaldean, and Aramaic. Columbus presumed Torres would be able to communicate with the Khan, because Marco Polo had written on his trips to China that Arabic was understood and spoken in that part of the continent. (Marco Polo had spent eighteen years in the grand court of the Great Khan in Mongolia and China.[32]) Columbus also sent Rodrigo de Jerez, a crewman who had experience dealing with a chieftain while exploring Guinea on the Coast of Africa.

The small party went along the Cocoyuguin River into the interior of Cuba hoping to find the legendary City of Cumbaluk, the Imperial City of Cathay and the home of the Great Khan. To their disappointment they found only a village of fifty palm-thatched huts. They were regally received. Torres and Jerez were offered ornate chairs to sit in that the natives called *duhos*. (Fig. C8) Ferdinand Columbus described these chairs as "made of one piece, in a strange shape, and almost like an animal which had short legs and arms and a tail, which is no less broad than the seat, lifted up for convenience to lean against, with a head in front and the eyes and ears of gold. These seats are called duchi." Jerez sat and enjoyed it, but Torres did not.[33] Torres called for a meeting with the chieftain, but was obviously disappointed that they had not found the

Great Khan, gold, cinnamon, or other spices. They soon decided to go back to Puerto Gibara where Columbus was waiting for them.

"While they were walking back through cornfields and along paths, the two 'cristianos' (Christians—Torres and Jeréz) met many men and women who were carrying glowing coals and good-smelling herbs in their hands. These were dried plants rolled in a large dry leaf. They looked like the small muskets made of paper that children played with during Easter festivities. The natives set one end on fire, and inhaled and "drank" the smoke from the other. It was said that in this way they became sleepy and drunk, but they also got rid of their tiredness. The people called these small muskets '*tabacos*.' "[34]

Fig. 8. Fr. Bartolomé de Las Casas, biographer of Columbus, historian of the New Indies and protector of the Indians. Museo del Hombre Dominicano, D.R. Photography by R. Oropeza 1993

They had not found gold or the Great Khan, but something that would be more important for centuries to come and would influence all cultures of the world—it was the beautiful *Herba Sancta*, nicotiana (tobacco). They would never have believed that within two centuries of this observation, the rest of the world, men and women alike, would have accepted this practice.

From Cuba, Columbus continued south until they found the island of Haiti,[36] as it was called by the Taíno. Columbus named it La Hispaniola, and it was there that the explorers made their first settlement,[37] the fort of La Navidad,[38] where thirty-five Spaniards[39] would be left for one year while Columbus returned to Spain. (The *Santa María* had sunk, so not all of the men could return to Spain in the remaining two boats.) Rodrigo de Jeréz, would take up the habit of smoking and later would be the first Spaniard to smoke tobacco in Ayamonte, Spain.

On December 29, 1492, the king of Haiti's brother met Columbus and took him to the best house in the village. Columbus was asked to sit in a beautifully carved low chair with a low back (a *duho*), which was offered only to dignitaries. Columbus asked about gold and was told that there was plenty in the Cibao (inland on Hispaniola).

On December 30 Columbus went to meet King Guacanagarí[40] and other *caciques*. The king took off his crown and gave it to Columbus, who in turn gave the king the cloak from his back and a ring.[41] On 13 January 1493 Columbus left for Spain.

In 1502 on his voyage to Veragua (Costa Rica), Ferdinand Columbus (son of Christopher Columbus, and his historian)[42] observed that "the chief men never ceased putting a dry herb into their mouths and chewing it, and sometimes they took a sort of powder [lime or other alkaline substance] they carried with the herb, which looks very odd." A similar observation was made by Amerigo Vespucci in 1499 on the island of Margarita, Venezuela.[43] This custom also was observed by Benzoni, (1565) who mentioned that the indians mixed tobacco with cockleshells[44] (cockleshells are alkaline, which releases nicotine, the active agent in tobacco). Gonzalo Fernández de Oviedo (or simply Oviedo) also made this observation while in Nicaragua, and he was able to differentiate somehow between the use of the hallucinogenic powder of piptadenia and tobacco powder.[45] Oviedo wrote about his experiences in his *Historia General y Natural de Las Indias*, which was published in 1526.

Benzoni mentions that "after the natives had finished their smoke, they became stupefied and they went to sleep in a strange sort of bed [*hammock*, a Taino word] which they hung up between the trees." Oviedo gave the following description of the habits of the Taíno: "The Indians have one that is especially harmful, the inhaling of a certain kind of smoke which they call tobacco in order to produce a state of stupor."[46] (Fig. 9) Oviedo goes on to describe the ways they smoked tobacco and

Fig. 9. Benzoni's depiction of stupor under the influence of tobacco smoke (1665). Indians smoked strong tobacco. NYPL.

the effects it had upon them. The Spaniards probably were incorrect in their interpretation of the effects of tobacco since the Taíno often mixed tobacco with hallucinogenics.[47]

In his *Historia General,* Oviedo refers to the custom of smoking cigars among the people of Nicoya (Nicaragua). Smoking was also observed in Coro (Venezuela) in 1526 by Federmann, who became a German governor of that land. Later on, in 1538, the Apure Valley in Venezuela became a famous tobacco center.[48]

Oviedo was instrumental in popularizing the word tobacco (*tabaco*).[49] Bartolomé de Las Casas also used this word in his publication of 1552.[50] The Spanish historian Gómara in his *General History of the Indies* in 1553 described the use of an herb by the Indians of present-day Colombia that they called " 'jop' and 'osca' and that they take the smoke." This was actually "*yopa*" (yopo) from the hallucinogenic P. Peregrina. It was called *hosca* in the Chibcha language of those natives. Tobacco was called *twaka, twa, aka, koggaba* or *choco* by the natives of Colombia, and in Ecuador it was called *cayaba*.[51]

Fray Bartolomé de Las Casas, called "Protector of the Indians," gave an account in 1527 of the custom among the people of Cuba and Hispaniola:

The herb which the Indians inhale is rolled like a sort of bundle in a dried leaf. . . . They then light one end of it and draw in the smoke at the other; the effect is a certain drowsiness of the whole body accompanied by a species of intoxication, in which state they declare that they no longer feel any sense of fatigue [strong nicotiana]. These mousquetons, or tobaccos, as they call them, have been adopted also by the settlers in this regions; I have seen many Spaniards in this Island of Hispaniola who use them and who, when reproached for such a disgusting habit, replied that they found it impossible to give it up. I cannot understand what enjoyment or advantage they derive from it.[52]

Las Casas also described the container in which they carried these leaves. This *calabash* was a gourd, and it was sacred to all cultures of the newly discovered continent. (Fig. 10)

Fig. 10. L. God Ixtac Mixonatel carrying the sacred calabash on his back
R. God Cinteotl with the calabash in the temple. Borgian Codex (16 c.) From Muriel G. Porter's Book of Mexica Culture Pipes.

Ramón Pané brought back to Spain some tobacco plants from Hispaniola after his second trip with Columbus in 1493.[53] The cultivation of tobacco at the beginning of the 16th century by the natives of Hispaniola was for both personal use and religious purposes. Diego Colon in his last will, dated May 2, 1523, mentioned that the planting of tobacco was increasing in Hispaniola. The Spanish settlers started cultivating tobacco for export on the Island of Hispaniola in 1531 after they had exhausted the gold mines by 1519.[55]

At the time of Sir Walter Raleigh's second exploration of the Orinoco in 1617–18, the Portuguese had begun cultivating tobacco for export on a large scale,[56] but they had been sending smaller amounts to Europe in the 16th century. The English settlers began exporting scented tobacco (n.tabacum) from Virginia in 1613 and from Maryland in 1631. The Real Audiencia of Spain, which dealt with the affairs of the new colonies, recognized in 1678 the importance of the success of the French cultivation of tobacco in the west of the island of Haiti. This production had increased from 20,000 quintals (a quintal is 100 pounds) exported to France and other parts of Europe, to one million pounds. Four thousand people were involved in the tobacco industry on the west coast of Hispaniola in 1677–78 under the protection of Louis XIV.[57]

Thirty years after the discovery of Hispaniola, the Taíno population of the Antilles became almost extinct. It was calculated by Bartolomé de Las Casas that the population of Hispaniola decreased from 400,000 or 600,000 in 1492, to only 4000 in 1512.[58] Eventually the Spaniards had to import labor from the Canary Islands and from Africa. The first black slaves were brought to Hispaniola in 1501, only nine years after the discovery of the new continent. In 1527, many slaves were brought to Hispaniola; in 1546 they imported 12,000 slaves. In 1568, 100,000 slaves were brought to the Caribbean islands to work in the sugar, tobacco and ginger plantations.[59] The Portuguese brought in the largest number of slaves to work in their colony of Brazil. Thus began the slave trade that would lead to an immense catastrophe in the years to come, not only in Hispaniola, but throughout the newly discovered lands of America.[6]

4

THE TAÍNO

Tobacco Use and the Ceremony of the Cohoba (Tobacco) in Hispaniola (Dominican Republic) in the 16th Century

One of the first reports of smoking by the Taíno appeared in the work of Fray Ramón Pané in his book *De Insularium Ritebus* in 1497.[61] Fray Pané traveled with Columbus on his second voyage to America in 1493 and was commissioned by the Spanish Pope Rodrigo Borgia, or Pope Alexander VI, to convert the natives of the New World to Christianity. Fray Pané described the festivity of the *cohoba*, wherein powder is inhaled through a Y-shaped, forked instrument made of clay or wood. (Fig. 11) The historian Peter Martyr at that time published Oviedo's observations that "they snuffed ground tobacco or *cogiabo*," probably confusing tobacco with the hallucinogenic piptadenia (identified by Alexander von Humboldt as Anadenanthera Peregrina).[62] Piptadenia is also known in the Orinoco basin as *yupa*, (yopo) which is mixed with tobacco and lime.[63]

Following is Pané's account of the use of tobacco by Tainos in 1497:

> The Indians have a certain kind of smoke which they call tobacco, in order to produce a state of stupor. The caciques employed a tube, shaped like a Y inserting the forked extremities in their nostrils and the tube itself in the lighted weed; in this way they would inhale the smoke until they became unconscious and lay sprawling on the earth like men in a drunken slumber. Those who could not procure the right sort of wood took their smoke through a hollow reed [*cañuela*], it is those that the Indians called tobacco, and not the weed nor its effects, as some have supposed. They prize this herb very highly and plant it in their orchards or on their farms for the

Fig. 11. Inhaling cohoba through a Y artifact and other paraphernalia. L. A calabash. R. Bowl to mix the cohoba. Museo Del Hombre Dominicano, D.R. Photography by R. Oropeza (1993)

> purpose mentioned above [Pané is describing the tobacco plant; the A. peregrina tree is a large. nine-foot tree in the mimosa family]. I cannot imagine what pleasure they derive from this practice, unless it be the drinking which invariably precedes the smoking. I am aware that some Christians have already adopted the habit [tobacco smoking], especially those who have contracted syphilis [caught from the natives] for they say that in the state of ecstasy caused by the smoke they no longer feel the pain, which they make their excuse for it is certain that smoking will never cure their disease. Lately to many of the Negros who live in this town, and indeed in all parts of the island [slaves recently had been brought from Africa to work in the mines and plantations], have acquired this habit. It seems to me that here we have a bad and pernicious custom.[64]

Bartolomé de Las Casas, referring to the Taíno Shamans during a festivity by the *Behequis* or *Boitios* (or *the Behequi priests*), described the effects of the "certain powder of dry leaves of brown cinnamon color": "The head remained on the side for a while, (Fig. 12) the arms placed on the knees and then raised the face towards heaven saying certain words, that must have been prayers to their gods (Anglería)."[65] These dry brown leaves were crushed and made into a fine powder in fine bowls (urns) made of guayacán wood of a dark color, often carved with a head of a god. This ceremony was practiced in *malocas* (temples) that were darkened, which favored or enhanced the hallucinogenic effects of the drugs during the ritual of the Cohoba.[66] (He is describing tobacco leaves.)

Fig. 12. Behequi (priest-shaman) under the influence of the cohoba, with his head inclined *to heaven*, to communicate and be possessed by the gods. In front of him there is a cemi (god or fetish). Museo Del Hombre Dominicano, D.R. Photography by R. Oropeza (1993)

Now we know that what they were smoking was either a different plant of tobacco or it was mixed with another hallucinogen. Father José Gumilla, in reference to the present Otomawa of the Orinoco, stated:

> This hallucinogenic plant is known as *anadenathera peregrina* [piptadenia] or *adenonthera peregrina*, which contains a hallucinogenic substance with a similar effect on the nervous system as LSD, causing visual hallucinations. This plant was used by priests or shamans called *behequis* or boitios while sitting in a *duho*, a special carved chair to obtain communication and possession by their gods. While under the effect of the hallucinations at least 10 different *fosfenos* were elicited with different geometric patterns.[67]

They saw birds that took their spirits to heaven, and they saw psychedelic colors that might have inspired the design of intricate patterns of fabrics, paintings and artifacts after the spell was over.[68] (Fig. 13) Wie

Fig. 13. A Huichol's visual hallucination under the influence of peyote woven in cloth by Mariano Valadéz from Los Altos de Jalisco, México. Photography by Mario Antonio Pacheco

Safford in 1916, and the Smithsonian Institute in the same year, supposedly made a clarification on the identity of the substance used in the festivity of the *cohoba*, but there still remains some doubt as to what the Taínos were inhaling.[69]

The behequis or boitios (shamans) used magic to cure the ill during special religious ceremonies. Their purification rites before this ceremony were similar to those of the Carib of the Otomacos and other Indians of the Orinoco-Amazon Basin and this ritual is still practiced up to the present time. They had to fast, drink tobacco juice and make themselves vomit before the ceremony, inserting a spatula in their throat to make themselves vomit.[70] According to Father José Gumílla, the Otomaco and Saliva of the Orinoco Basin inhale yupa [piptadenia] mixed with tobacco and lime during their ceremonies and when they were getting ready for war."

In the first quarter of the 16th century, when the Tainos were almost completely exterminated and had been replaced by African slaves,[71] some of the Taino rituals using tobacco were adopted by the Africans, who continue to use tobacco and "holy blue water," still used to this day during voodoo ceremonies. Tobacco was also used by the Taíno to conjure up evil spirits from the sick persons by blowing smoke on them. This ritual is still practiced in voodoo ceremonies in the Dominican Republic and Haiti.[72]

In 1565, ninety-two years after the discovery of the new world, a Milanese traveler, Girolamo Benzoni, traveled for fourteen years through South and Central America. Benzoni wrote about tobacco:

> When the leaves are ripe they are picked up from the bush. They tied them in bundles and they hung them close to the fire till they are dried and when they want to use one they take a leaf from their grain [corn] and roll the tobacco inside of this leaf and start fire in one side and smoke it through the other end. The smoke then penetrates the mouth, the throat and the head. They keep the smoke in as long as they can until they lose their senses. Some of them fall to the ground as if they were dead and remain that way most of the day and night. . . . Others get happy only when they get dizzy. You can see how abhorrant and how poisonous from the Devil, this is.[73] (see fig. 9)

Benzoni may have been wrong in ascribing these effects to tobacco only, as had Las Casas and Oviedo.[74] The effect may have been caused by mixing the hallucinogenic peregrina with tobacco. On the other hand, n.rustica has a much higher content of nicotine than the present n.tabacum. This strong tobacco might have given similar psychotropic or hallucinogenic effects as the powder of piptadenia.[75]

Thevet (1555–1557) gave a description of how the people of Brazil used the tobacco smoke to cure some illnesses, and how the witch doctor (shaman) smoked the cigar and blew the smoke on the face of the sick. (see fig. 5) This practice still exists in some parts of the Dominican Republic and in other countries like México. Benzoni wrote in 1565:

> See what a pestiferous and wicked poison from the devil this must be. It happened several times to me that, going through the provinces of Guatemala and Nicaragua, I have entered the house of an Indian who had taken his herb, which in the Mexican language is called tobacco, and immediately, perceiving the sharp, fetid smell, I was obligated to go away in haste and seek some other place.

One of the first observations regarding addiction to tobacco was made in 1515 by Fray Bartolomé de Las Casas, "a Spaniard, married and honest, that took tobacco and the smoke from it, in the same way of the natives and he said that because of the great pleasure that he obtained from it he would not give it up for anything." This man obviously was already addicted.[76] Later, Francisco de los Cobos, a Franciscan, mentioned how the Spaniards became addicted to tobacco from the Taíno caciques.

The African slaves also adopted this habit, to "relieve their tiredness" according to Benzoni.[77] Oviedo in 1546 also mentioned that many of the slaves sowed tobacco on the properties of their masters and smoked it after they had finished their work to relieve their tiredness.[78] (Fig. 14)

The Spanish clergy in Santo Domingo at the beginning of the 16th century considered smoking a violation of the christian religion because smoking was a part of the religious ritual of the Taíno, and thus considered a heresy. This was the first official condemnation of tobacco by an institution on religious grounds, as far as we can determine. It was followed by the Edicts of Lima (1583) and México (1575, 1585).[79] About the same time Fray Tomás Ramón stated that those who smoke or take tobacco powder (snuff) "have some pact with the devil."

Bernabé Cobo (1582–1657), a Spanish missionary in Hispaniola, referred to the smoking habits of the natives and early settlers: "The Spaniards took the habit [smoking] from the Indians of Hispaniola. The caciques [chiefs] used to take it in the following way: they put dry leaves, after being cured, in tubular pipes curiously carved. They would light it on one side and suck it through the other drinking [in the 16th century smoking was referred to as "drinking"] the smoke."[80] Francisco Hernández de Toledo, a botanist sent by King Felipe II of Spain to study plants in México, stated: "They call them *tauacos* in Hispaniola, two segments

Fig. 14. Chartoyer. 'A Black Slave and His Wives.' St. Vincent Island. Black slaves took to the practice of tobacco smoking soon after being brought to the American continent. The Scourge of the Indies. Maurice Bensson. Book of Buccaneers (17th c.)

of hollow reeds a hand and a half long. Outside they are smeared with soot and inside they are full of tauaco [*tauaco* is "tobacco" in Taino]." Taíno usually smoked tobacco in cigars, which explains why very few pipes have been found in Hispaniola. (Dominican Rep.—Haiti)

*****

Almost immediately after the discovery of America, the European settlers became abusive of the natives. The settlers Columbus left behind in the Fort of Navidad in Hispaniola following his first voyage forced men to work and raped indigenous women. Some of these women got pregnant and committed suicide either by drowning in the ocean, hanging themselves from trees, or eating unprepared cassava (yucca, which is very poisonous in its raw state). When Columbus returned to Hispaniola on January 16, 1493, one year after his return to Spain, he found that the settlers had been murdered.[81]

The "white gods" had become settlers, conquerors and enslavers of the newly discovered continent.

The Taíno are said to have given their colonizers two gifts—syphilis and the habit of smoking—which were then propagated to Europe and the rest of the world. We might call this "the Taino revenge."

The Buccaneers. Antilles Islands and the Pacific Islands.

Tobacco use was spread through the Caribbean Islands and coasts of Central and South America in the 17th C by French and English soldiers that had served in the Thirty Years War in Europe 1618–1648 and had no other means of survival except for looting the coasts of these islands or looting ships transporting gold, silver, sugar, tobacco, or any other merchandise. They were called bucaneers or pirates. Some of these pirates were legendary: Morgan, Dampier, Anson, Lafayette, Robinson Crusoe (Alexander Serkirk 1719) and others. (Fig. 15)

Fig. 15. A buccaneer with a lighted pipe. The Scourge of the Indies. Maurice Bensson 17th c

5.

CULTURES OF MESOAMERICA

Cultures of the Gulf: Olmec-Totonac
Olmec 1000 BC–1 AD
Totonac 1 BC–7 AD

"Mexicans be aware!
These are the roots of your history and your pride.
You will hear the silent voice of the oldest culture
of México, the oldest civilization in America.
The Olmecs (then Totonacs), transformed rains in
harvests, the sun in calendars, stones
in sculptures, cotton in textiles, journeys
in commerce, mounds in thrones, jaguars
in religion and men in gods."
Augustín Acosta Lagunes
Museo de Xalapa Ver. México
Translated by R. Oropeza, M.D.

The Olmec is the most ancient civilization in the Americas, dating back to 2000 B.C. (Fig. 23) This culture developed in the areas of present-day Veracruz-Tabasco, México, and had a large influence on the future development of the rest of the cultures of Meso, North and South America. The Totonac are descendent of this culture. (Fig. C11)

The first European explorers to the Coast of Yucatán and the Gulf of México were sent by Governor Diego Velázquez of Cuba (1517–1518) looking for, among other things, slaves to work in the new sugar plantations of Cuba.

The Totonac cultivated tobacco extensively in the present states of Veracrúz and Tabasco where two cultures, the Totonac and the Maya, met. Tobacco is still cultivated extensively in the area of Santiago and San Andres Tuxtla, where there is a large cigar factory. Most pipes are dated back to the 7th—8th centuries. The shamans of this and other cultures, like the Maya, chewed tobacco during their healing and/or religious ceremonies. According to Mary Miller "there were early classic depictions of God L—suggesting Maya interest in the tobacco trade before AD 550." (Miller, personal communication) Spinden-Schele Coe)

The offering urn (Fig. C12) a figurine of unknown proveniecne depicts a shaman chewing a roll of tobacco. He has a sacred bundle hanging from his neck containing 8 rolls of tobacco. The back of this urn exhibits an opening where the tobacco was inserted and burned during religious ceremonies. At the top is an opening for the smoke to go to *heaven*.

The second figurine also of unknown provenance depicts a warrior smoking a large cigar (Fig. C13). This most likely is the oldest proof of cigar smoking by an indigenous American. It was found by the author in the museum Rufino Tamayo in Oaxaca, México. These two figurines have been dated to about the 6th-8th centuries.

Another figurine at the museum of Xalapa is also an offering urn to burn tobacco. It depicts a man with *ascites* ("fluid in the belly"), considered by the Totonacs to be a sacred man. This figurine was found in the archeological excavations in Veracruz.

The Maya (500 BC–1535 AD Classic Period 2 AD–8 AD)

Francisco Hernández de Córdoba (1517) and Juan Grijalva (1518)
First Explorers of the Southeast Coast of México (Yucatán-Campeche, Tabasco and Veracrúz) (Fig. 26)

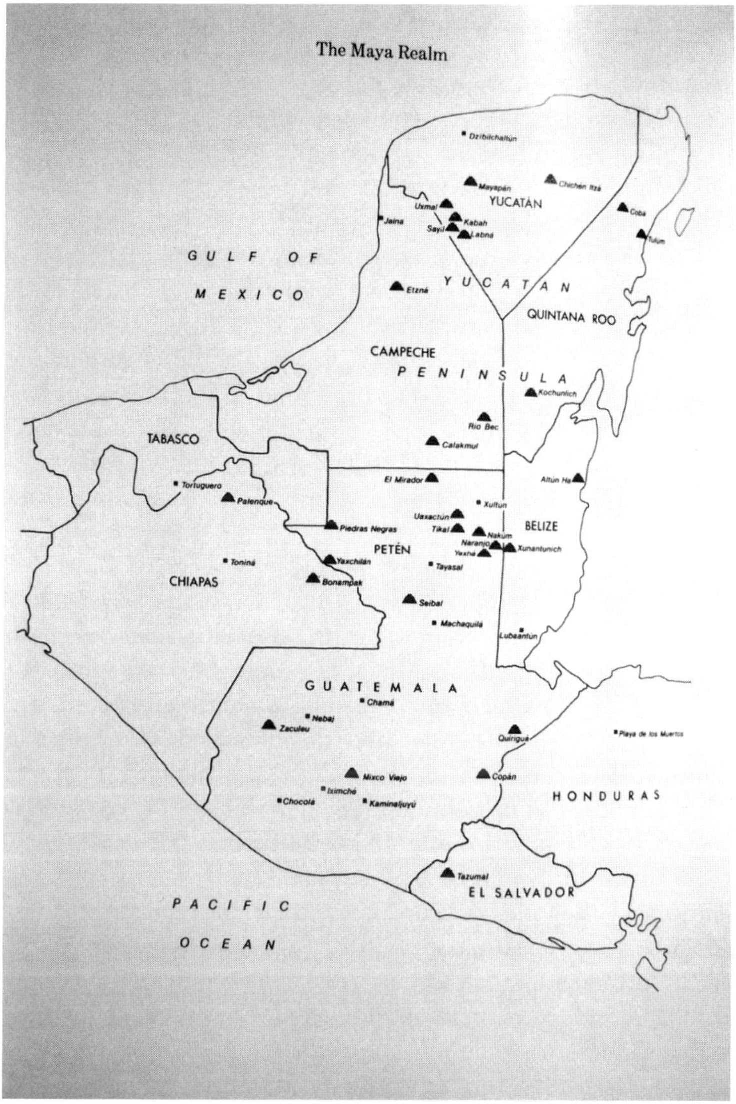

Fig. 16. Map of *The Maya Realm.* University of Oklahoma Press.

A Maya Dawn Prayer

"Look at us, Hear us!
Heart of heaven, heaven and earth!
Give us our descendents, our succession
As long as the sun shall move

And there shall be light, let it dawn,
Let the day come!
May the people have peace, much peace
And they may be happy and give us good light
Let there be dawn, and let the light come!''
From the Popol Vuh, 1555 A.D., (Maya)

The Maya is one of the oldest civilizations, dating back more than 500 years. It extended from the present southern part of México to Central America. The Maya were ruled by Kings; they built pyramids to their gods and practiced sacrificial rituals. They had a god of commerce, God L,[82] precursor to 1 or 7 Deaths (Fig. 17) found in the 19th century in the ruins of the building of the Cross in Palenque, Chiapas, México. (Fig. C14) Palenque was the splendid center of the Maya civilization in the 7th—8th centuries, the classic Maya period, with a writing system of 500 signs.[83]

According to Bishop Landa, who came to convert the Maya to Christianity in the beginning of the 16th century, Maya, like other cultures in the Americas, had plantations of food crops as well as cotton and tobacco. Bishop Landa related that the Maya, like most Mesoamerican civilizations, made fermented alcoholic beverages using maize (corn) and the juice of the *agave maguey* (cactus), and a drink called *balche*, used in religious rituals, made from fermented honey and the bark of the balche tree.[84] During religious ceremonies they also smoked the rolled leaves of the tobacco plant, which were more potent than today's domestic varieties and induced in them a trance-like state.[85]

Francisco Hernández de Córdoba sailed from Cuba on a slave-gathering expedition on February 8, 1517. He was blown off course and landed three weeks later in Cape Catoche, Yucatán. Córdoba was surprised to find well constructed buildings and a militaristic people who engaged him in war. He suffered multiple wounds from which he did not recover. He passed through present Campeche on his way back to Cuba. During this expedition he was offered lighted cigars to smoke by some of the Indians that he encountered.[86] Francisco Montejo went on the 1517 expedition with Córdoba. While in the Isla de los Sacrificios (Veracrúz), Montejo was offered black reeds *cañutos* (black, scented pipes-)with tobacco.[87]

Juan Díaz, chaplain of Juan Grijalva,[88] mentions that Grijalva landed on the island of Cozumel in 1518 during an expedition from Cuba to survey the coasts of the new lands of America. Oviedo reported that

Fig. 17. God L of Palenque Chiapas. God of commerce, trade and tobacco, represented at his peak on the Murals of Cacaxtla Tlaxcala Mexico. The precursor to God one and seven Deaths of the Popol Vuh. He is a toothless old man, but wicked and ruthless with a jaguar skin on his back. Smoking a long tubular-tobacco pipe, his best advertisment is his product. Wall panel at the Temple of the Cross. Palenque, Chiapas México.

when Grijalva first made contact with the Maya in 1518, a chief "gave the general and to each of the Spaniards . . . a little hollow tube burning at one end, made in such a manner that after being lighted they burn themselves out without causing flame, as do the incense sticks of Valencia . . . and they smell of fragrant odor . . . they tied their cigars with a string of hemp.[89] (Fig. 18) The Indians made signs to the Spaniards not to allow that smoke to be lost [that is, they should inhale it] just as if they were taking tobacco."[90]

Bernal Díaz del Castillo,[91] who in 1519 would accompany Hernán Cortés during the Conquest of México, described this custom of the Toto-

Fig. 18. Plate (ceramic) of Uaxactum. Guatemala. Probably a priest sitting in the lotus position smoking a large cigar. From Muriel G. Porter's Book of Pipes.

nac and Maya on a previous trip in 1518. He stated that the Spaniards were "fumigated with incense from the Maya Priests." Díaz made this observation of the local people while on the Tabasco River (Villahermosa, Tabasco, México): "On the following day more than thirty Indians came to the headland of the palm trees where we were encamped, bringing roasted fish and fowls, zapote (custard apple) fruit and maize cakes, also braziers with live coals and incense, with which they perfumed us all."[92]

Some of the Maya chewed *moy* (or *moi*, tobacco powder with lime) every day. Among their superstitions was the belief that if they kept a wad of tobacco in their cheek when it thundered, the thunder would not be so loud. They also believed that when they died, this *pilico* (moy) would protect them while they rested.

The Maya believed that the life of a person started at the moment of conception. The priests consulted their horoscopes as to the positions of the planets and at that particular moment they would give a name to the child in order to give a name to the child. According to the sacred writings of the Popol Vuh (Codices of the Maya), the gods created and gave form to humans. The gods made the four first humans of *masa*, and their spirit was given to them through tobacco smoke.[93]

"Reed cigarettes" (tubular pipes) as well as cigars were commonly used by the Maya and other peoples of Mesoamerica during religious and other ceremonies, including the ritual of puberty (also called the rite of passage, or EM KU, meaning "coming of god") as described by Bishop Landa.[94] Diego de Landa, a christian friar, described this custom in detail[95] and remarked on its importance. "Baptism: they had such devotion to it that all of them received it and they had it in great reverence. Those who had sins had to confess them to the priests and they had so much faith on them that they did not commit them again."

The Maya hung a white round bead attached to the heads of their male children over the age of three years.[96] By doing so, the boys would become cross-eyed, a sign of distinction. The girls had a shell that would cover their private parts attached to a string tied around their waist. It was sinful (Landa's interpretation) and ugly to remove these two objects before the time of baptism at the age of twelve. They never married before this time.

On the day of the baptism, they would bring all of the children that were going to be baptized to the house of one of the parents, who was giving the party. They placed the boys on one side and the girls on the other side. This was done in the courtyard of the house, which was kept very clean and covered with fresh leaves. The girls were placed under the care of an old woman, and the boys were under the care of a man.[97]

Once this was done the priest (*chilan*) had to purify the courtyard to conjure up the devil. (Landas' christian interpretation)

The *chaaks* (wise men) were placed in each corner of the patio. They would sit holding a rope forming a square, with the children in the square.[98] Landa described what followed:

> The priest then would tell them to be quiet and he would start giving the grace, with his "*hispo*" (sacred stick) to all of them. He would then give the father who was giving the party a bone and with this he would touch each child on the forehead. He then put "holy" water on their faces and in between their fingers and toes without uttering a word. This water was

made from certain flowers, cacao mixed with "virgin" water that they said they have brought from the hollows of trees or from the stones of certain mountains. After finishing rubbing this water, the priest would stand up and remove the white cloth that they had on their heads and other clothes with beautiful color feathers and cacao pods. This was gathered by the chaces (old wise men) and then the priest would cut the string from the forehead where the bead was hanging in between their eyes (boys). Behind him the priest's assistants carried a bouquet of flowers and a cigar (*humazo*) that the Indians have a habit of smoking (*chupar*) and they were touched nine times with this bundle, making them smell the flowers and to smoke the cigar (*chupar el humazo*).[99] The priest (*Chilan*) purified them with copal (incense) and tobacco smoke. They made an arduous trip to the Island of Cozumel to worship at the shrine of the Goddess Ixchel, the Goddess of pregnancy and life, as part of the ceremony they smoked tobacco.[100]

The person who gave the party had fasted for three days and would fast for nine more days, "and they always complied with that ritual," according to Landa.

Among the K'iche Maya (Guatemala) tobacco had a magic healing power[101] and played an important role in their religion,[102] as we can see in the myths and representations of their codices. One myth held that the gods of the four winds were great smokers. Falling stars and lightning were the loose sparks of their big cigars. The Madrid Codex depicts priests squatting or lying down, smoking and sending the smoke to heaven. The smoke and sparks, according to Wolf, is a magic ceremony to elicit rain.[103] (Fig. 19)

The Maya goddess of medicine, Ixchell was also the goddess of life and death. (Fig. 20)[104] The Maya offered, among other things, tobacco to which they attributed medicinal and divine properties. During these festivities they also drank *balche*, a drink that was made from the bark of the balche tree, and fermented honey which they considered to have been created by god.[105] (Fig. 21) (Fig. C15)

In this culture, as in any other culture of the past, myths and legends were important in their daily lives. (Fig. 33a) A K'iche legend in the Popol Vuh tells us that Hun-Camé and Vuqub-Camé,[106] or 1 and 7 Death, gods of the underworld, gave a test of endurance and worthiness to the legendary Twin Brothers Hun Hunahpu and Vuqub Hunahpu and later to Idunahpu and Xbalanke who have to overcome the legendary darkness of the underworld as a test of their worthiness to play in the ballgame called Pitz by which communities often settled their differences rather than by going to war. The captain of the losing team usually was decapitated. This custom extended throughout Mesoamerica and the few cultures of

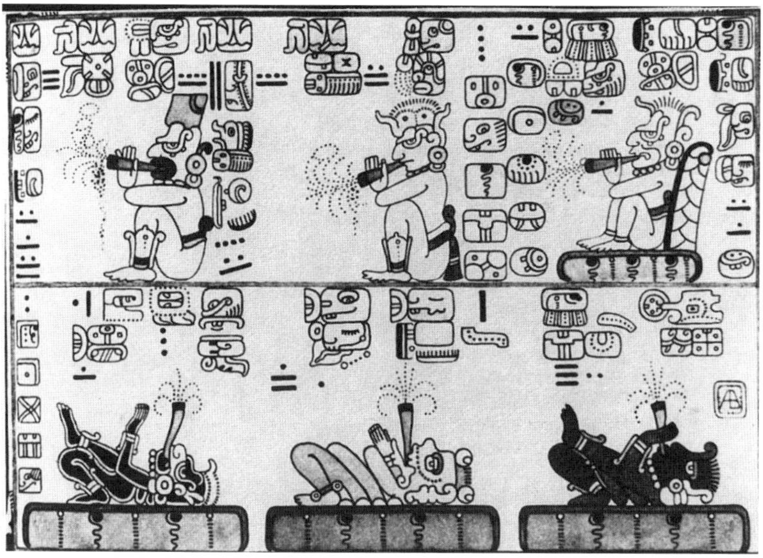

Fig. 19. 'The eternal joy of the gods.' 'The Maya gods were great smokers.' Priests smoking cigars to the heavens, the sparks symbolize the stars. Madrid Codex. N.Y.P.L.

the present-day United States where there was Maya (or Toltec) influence. According to the Historian Archeologist Mary Miller in her book *Blood of Kings* (Schele-Miller) there was a strong relationship between warfare and the ball game, called by the Maya Pitz. They have tricked the gods by wits and deception. (Personal communication)

The Twin Brothers went to Xibalba, where they were supposed to compete in a ball game. They had seven days before the ball game started, and had to go through tests of endurance. If they passed the tests, they would be fit to play. Before they had left their home they had told their grandmother that they had planted a cane in the center of the garden. If the cane died, that meant they were dead. But if the cane flourished, they were alive and would be back.

They arrived at the dark house of the underground, where they were given burning firewood and a lit cigar. The light was supposed to last only through the next morning. The people who were putting them to this test were the gods of the underworld, Hun Came and Vuqub Came who knew that through this test it was impossible to survive and in this way they would surely be killed. The twin brothers understood the treachery of this test, but they figured out how to get out of it. They used

Fig. 20. Goddess Ixchel. A dual goddess of goodness and evil, goddess of the floods, death, birth, life and weaving. The owl, top and bottom, symbolized death; a floral medallion at the center of the chest symbolizes that corn is sprouting (life), it is the maize god (one maize). the newborn between her legs at birth symbolizes the beginning of life. Photography and private collection: R. Oropeza

Fig. 21. Descending God of Tulum. He brought white honey and fermented honey (mead) to earth. During celebrations to this god, balche and tobacco were shared by the participants. Tulum Q. R. México. Photography at the ruins of Tulum by R. Oropeza. 2004

the red feathers of the macao bird to make it look like fire, and put fireflies at the end of the cigar to make it look as if it were lighted. When the guards came in the next morning sure that the fire had died out and the cigar had burned away, to their surprise they found that the fire was still burning. Hun Hunahpu and Vuqub Hunahpu had started the fire just before the guards entered. The guards could not explain how the fire had lasted so long. They thought that the Twin Brothers had supernatural powers, and judged them fit to play and "after conquering hell itself, rose up to the sky to become heavenly bodies."[107] (Fig. C16)

The Lacandon Maya

"Here is my tobacco
I give it to you O Lord
I'll smoke this cigar
O Lord, I pour, bring it

to your mouth.
My container of balche
A bit of this liquid is my gift."
(translation by R. Oropeza from Demetrio Sodi)

"Kootz (tobacco) is naturally healing and so medicinal that there is hardly a disease to which it is not applied."
On describing tobacco's medicinal properties.
Ricardo Osado, XVIII C. Libro del Judío. (Book of the Jew)

Palenque (Chiapas, México) was an important Maya center of culture in the 8th century. The Lacandon Maya inhabit the Selva Lacandona, a national preserve of the rain forest in Chiapas, México. They are still living in that jungle today,[108] and cultivate the tall tobacco plants (nicotiana tabacum) to make their cigars in the way they have done for centuries. These unconquered independent natives of the jungle fought Hernán Cortés in the 16th century and they are still fighting the present government of México through their leader, Comandante Marcos (though he is not a Lacandon). They are the only Indians of America that remain faithful to their traditions, among which is the cultivation and use of tobacco. Through their present unchanged ways we can infer how their ancestors lived and how they smoked tobacco in cigars. It is likely that they started the cultivation of wild tobacco (such as *nicotiana sylvestris* and *n.otophera*) long before the classic Maya period in the 7th century A.D. (Miller personal communication.) Tobacco was called *Zig, Kuts* and *Mai* by the Maya.[109] They sowed tobacco seeds in large pots or small plots. The sprouts were then transplanted to the fields in October. The Maya celebrated religious rituals on this occasion, like many other Amerindians did before Columbus. They dried tobacco in the sun and hung the leaves in special sheds for two weeks. Then they pressed the leaves on flat stones, on and off, until they ceased to produce heat. They prepared large bundles, not only for their personal use, but also for barter or sale. These were among the first observations of the Lacandons by Bishop Fray Bartolomé De Las Casas in the 16th century.[110]

Fray Bernardino Sahagún also gave an account of Lacandon customs in the 16th century: "The seller of fine tobacco, sells fine tobacco . . . small tobacco. He rubs it between his hands. He who finds fine tobacco chews it and some prefer wormwood tobacco. Fine tobacco affects one; it makes one drunk, it aids one's digestion, it dispels one's fatigue."

Francisco de Paula García de Peláez, a Guatemalan historian in 1857, described what the Lacandon Indians left behind while fleeing through the jungle during a war with the Spaniards: "utensils made of stone, knitting needles, cooking and eating ware, tobacco twisted into the leaves of the nance tree which they called '*puquiet*.' " These Indians, he said, lit their tobacco by "rubbing a cork attached to a small mill."[111] They pulverized tobacco sometimes and mixed it with a gruel of sweet potato and corn, and kept it in a gourd. More recently, the Chorti Maya added *aguardiente* (60 percent alcohol) with molasses and anise, rather than making *cocom* as described above. (Xtabentún)

According to the historian of Maya culture, Mary Miller, in her book *Courtly Art of the Ancient Maya* (pages 51–52) one of the supernatural characters of Maya worship was God L considered the 'patron of trade, tribute and commerce' but also one of the princes who rules in the Darkness of the Underworld, depicted at the wall of the Temple of the Cross smoking a large tubular cigar.[112]

The Tzotzil and Lacandon cultivate a hybrid variety of nicotiana tabacum that grows very large leaves from which they make their giant cigars. (Fig. 22) They also use this tobacco after it has been ground and mixed with cane liquor "for treating diseases of the prostate and to alleviate urinary problems."[113]

The Tzotzil and Lacandon believed in the power of tobacco to protect them against the gods of the underworld. They would put a wad of chewing tobacco (*mui*) in their mouths against their cheek, and this would give them complete assurance of protection against these evil spirits.

The main source of information on the habit of smoking by the Maya prior to the discovery of the Americas comes from the Popol Vuh (written in K'iche from Chichicastenango, in present Guatemala).[114]

Tobacco was chewed by the shamans in Huehuetenango (now Chilcastenango), Guatemala.[115] From the scanty information we have, we know that they smoked mainly cigars and chewed tobacco during religious ceremonies. The Chaaks "gods of rain" enjoyed smoking, as we can see in the Madrid Codex where there are three examples of priests smoking tobacco. According to their legend, the gods of the four winds are great smokers and the falling stars are the falling ashes of their cigars. When there is lightning and thunder the gods are getting sparks by hitting rocks together so they can light their cigars. The descendents of these Maya natives still celebrate these traditions.[116]

The Chilam Balam book of the Maya of Yucatán contains a series of interpretations that relate to the use of tobacco (cigar):

Fig. 22. Pedro Kayum, 1958. A Lacandon Maya smoking a large cigar as his ancestors did more than one thousand years before. Hacienda Na Bolom. San Cristobal de Las Casas. Chiapas, México. Photography by R. Oropeza taken from the orginal photograph by Gertrude Duby Blom in San Gristobal de Las Casa.

The father asks the son, "Bring me the fire bug of the night. The one that gives scent from the north to west. That the tongue of the tiger comes with it." The son replies, "It will be done that way, father."

The explanation is that the father is asking for cigars. The tongue of the tiger is the fire to light the cigar.[117]

The Maya, like other cultures of Mesoamerica (México, Guatemala), also used tobacco for medicinal purposes.[118] They used tobacco in rubber enemas to treat constipation. (Other intoxicants were administered through rubber enemas; in Perú they use enemas of coca.)[119] (See fig. C4) The Maya of Guatemala treated infections and colds with a concoction made of snake powder mixed with fat, applied hot and wrapped in tobacco leaves. The shamans, or *curanderos*, still use tobacco during their healing ceremonies.

Dr. Francis Robicsek, in his book *The Smoking Gods*, describes the use of tobacco and presents a comprehensive depiction of surviving artifacts of the classic Maya period, stressing the importance of tobacco in this culture.

The Maya also smoked reed pipes filled with tobacco, and they smoked cigarettes. They pulverized tobacco and placed it inside of a rolled leaf of zapotille that they called *cocoom*.[120] Cocom was prepared by mixing the ashes of the tree cocom with mead (fermented white honey). Papaya sap was used as a binder. (Fig. C17)

Mary Miller, in her book, *Courtly Art of Ancient Maya*, 2004, illustrates the practice of smoking by the Maya of present Mexico and central America through ceramic vases, plates, and other artifacts found in those regions.

Central Cultures
Mexica (Aztec)

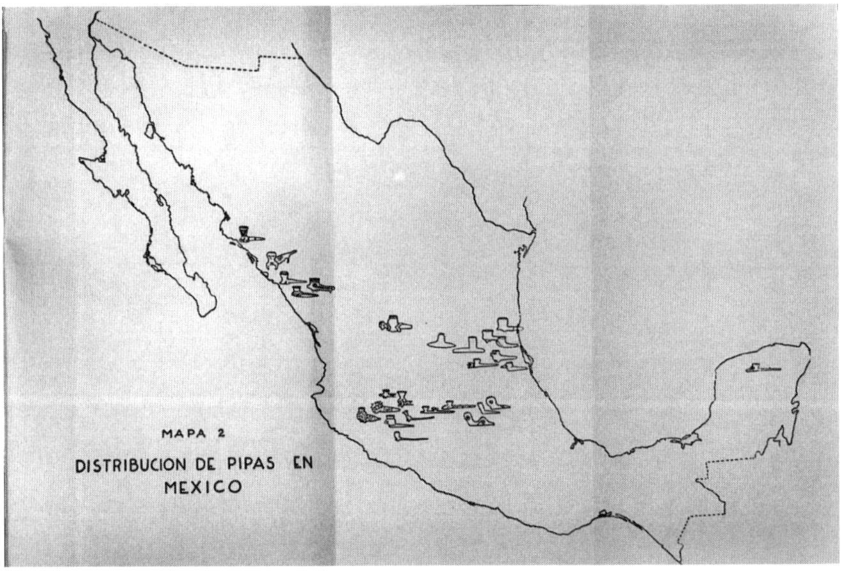

Fig. 23. Map of Pipes of México. Muriel G. Porter

"Rise Play the Drums"

"Rise proudly to the sound of drums
and let your friendship be known,
bare your souls to one another
for we are here on borrowed time.

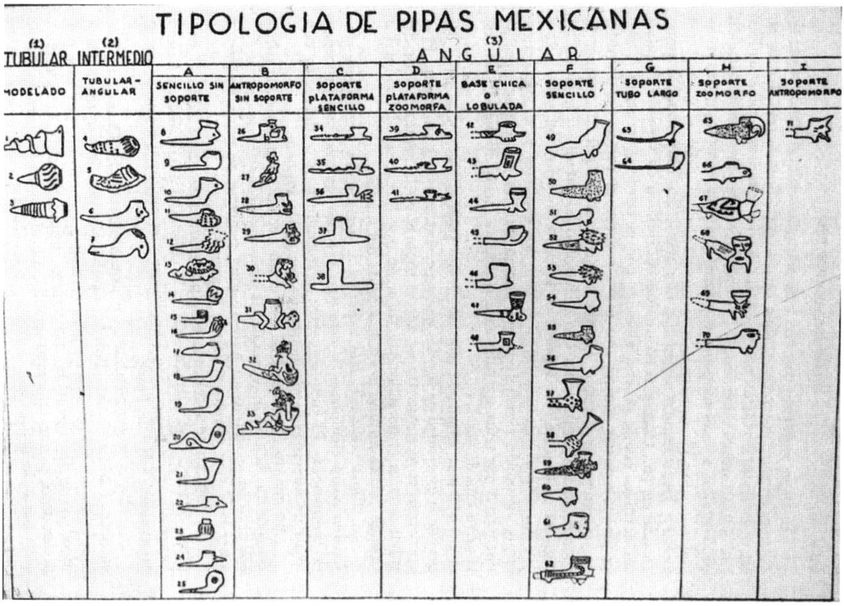

Fig. 24. Typology. Pipes of México. Muriel G. Porter.

We offer our sacred tobacco pipes-
Our sacrifice."
Cantos de Nezahualcoyotl (1431–72)
"Hungry Coyote"
King of Texcoco (México)
(Translated by Joyce Oropeza from José Luis Martínez, page 177)

The Aztec. Mexica Culture (1352–1521) AD

The Aztec culture (México) and the Inca culture (Perú-Bolivia) were the most advanced civilizations of the "New Continent" at the time of Columbus' discovery of America. The Aztec had a society ruled by Kings where sacrifices were continuously given to their main god, Huitzilopochtli, and other deities to appease them and to obtain favors from them in times of war or difficulties.[121] The Nahuatl language of the Aztec had a symbolic writing. They had a precise calendar of 365 days that did not have to be corrected for 52 years, and a religious calendar of 260 days that interlocked with the previous one. They had a vigesimal system of counting by using dots. They were advanced in astronomy like the Maya, knowing the changing position of the stars in relation to the Earth and the changes of the position of the moon, as well as recognizing the morning star (Venus) and other planets. Unlike the Inca, which had a communal society, the Aztec had freedom of commerce to sell their own products in markets through barter or using cacao beans as a currency. They also use T shaped pieces of zinc or copper.

The Aztec were warriors par excellence and dominated all of the neighboring communities and cultures, with which they were in a constant state of war. They waged special wars, called "flower wars," in order to obtain prisoners for their sacrifices. They never conquered the Tarascan culture of the west, however. We know about their daily life, customs, feelings and philosophy through their poetry. And we also know about the sophistication of this culture from the artifacts collected at the Museum of Anthropology of México City and through the writings of Sahagún. Tobacco was well known to them.

We have also obtained information about the use of tobacco by these cultures from the writings of other historians, among them Don Hernando Alvarado Tezozomoc (or simply Tezozomoc). He was the son of ex-Emperor Cuitlahuac, and was baptized and educated so he gave us information about the customs of Aztec. (Some royal Aztecs were educated in the school of Tlatelolco.) This firsthand information tells us how important tobacco was in Aztec religious rituals. We also learn from Tezozomoc that the shell of the squash or gourd where they kept their *picietl* (tobacco) was sacred to them. It was an important religious symbol, and also was used as a sign of command and power, like a scepter.[122]

Tezozomoc tells us that during Motecuhzoma's inauguration as king, Motecuhzoma was adorned with precious feathers and given a calabash to keep his piciete (tobacco), which was hung around his neck.[123] He also related the story of Motecuhzoma when he had gone into battle against the people of Tututepec and Quetzaltepec. The Aztecs had to make a long climb around the neighboring mountains of México to get to these towns. Tezozomoc described the ferocious battle that ensued, with many people dying on both sides. After having a rest following the battle, and before departing to Tenochtitlan (México City), Motecuhzoma hung his *calabacillo* (calabash) with his *picietl* (tobacco) as a symbol that he was in command and as a sign of departure. It also symbolized that he was an old chief and sage (although he was not old).

Before entering Tenochtitlan, Motecuhzoma was rubbed with an ointment prepared from large worms from "the trees that had spikes" (raspozos), which gave an abundant yellowish fluid when smashed. (These worms still exist in the trees close to México City and are called *chinagües*.) The Aztec considered this ointment of the worm *cuapatli* (*cocus axies*) to be a muscle relaxant. It was applied after long journeys and to relieve muscular pains. They also chewed tobacco mixed with lime (*tenchiete*) in a 10:1 ratio. They called this ointment *betun-axin* and used it to treat their wounds, mixing it with resin or the sap of the black pine (*ocote*) to make a paste. This paste was used to treat 'abscesses'. Tezozomoc mentions that he used it on himself to treat a ''hernia' (he probably meant an abscess or a boil), and he claimed he got relief from it.[124]

When the Aztec passed through friendly territories, they were given food and gifts, among them *perfumadores* and on one occasion *cañutitos* (small reeds). These could be interpreted as being incense urns (censers) and / or pipes.

After another battle against the people of Huexotzingo in which they again suffered large losses and captured many of the enemy, the Aztec returned to Tenochtitlan where Motecuhzoma and his commanders were given a large reception with cacao flowers and perfumadores (pipes) to smoke. This was followed by the sacrifices of the prisoners captured in the battle. Each soldier had to sacrifice his own prisoner, but before doing so, he had to engage him in a battle with a macuahuitl (sword). The prisoner was put on a large pedestal or large stone, called Tlacaxipehualixtli Tlahuahuana, (Fig. 25) and one of his legs was sometimes tied to a stone.[125] The prisoner rarely won the fight, and after being killed his chest was quickly opened while his heart was still beating. His captor or a priest (chalchiutepehua) ran to the top of the temple of the god Huitzilopochtli, who through this ritual sacrifice would give the warrior strength in future battles. Tobacco reeds were smoked afterwards. (Fig. C18)

Observation of Tobacco Use during Hernán Cortés' Conquest of the Aztec Empire

The ''after dinner smoke'' of Motecuhzoma.

''They also placed three gilded and ornate pipes (cañutos) containing liquid ambar mixed with some herbs that they called tobacco.''
Bernal Díaz del Castillo (1519)[127]

At the time that Cortés lived in Cuba, smoking by the Spaniards must have been infrequent. They were surprised and they thought it was a novelty to see Motecuhzoma smoking at the banquet. Bernal Díaz del Castillo, one of the soldiers of Hernán Cortés, wrote the Conquest of

Fig. 25. Sacrificial Stone of Tizoc or Cuauhxicali. (Now at the Ntl. Museum of Anthropology in Mexico City
It depicts a sacrificial ceremony to the god Huitzilopochtli, Tlaloc or other minor gods. The captive gladiators were tied to this stone to fight their captors for their freedom. México a Travéz de los Siglos. By Riva Palacio, 1885. Photography by R. Oropeza.

México 65 years later to refute the inaccuracies of Francisco López de Gúmara, secretary to Cortés.

> After their royal appetite was appeased, water was handed to him [Motecuhzoma], in the same manner as has been done before commencing the meal, for the Aztecs were as constant in their ablutions at these times as any nation of the East. Pipes [acayetl[126] were then brought, made of varnished and richly gilt wood, from which he inhaled, sometimes through the nose and others through the mouth, the fumes of an intoxicating weed called "tobacco,"[127] mingled with liquidamabar[128] and powder of charcoal.[129]

Cortés sent some tobacco plants to Charles V of Spain as a gift soon after his conquest of México City.[130]

We know from the Madrid Codex, translated by The Abate Brasseur de Bourbourg (1861), which recorded the customs and habits of the natives of México before the conquest, that the cultivation of tobacco was widespread. We also know that smoking was common not only by priests, but by the common people, because of the finding of pipes and ceramics (map 2) in different parts of México. The majority of these pipes were not ornate, but some were highly polished and carved with engravings for special occasions. Tobacco and artifacts to hold the pipes were sold in the market of Tlatelolco (México City) by official vendors. Home grown tobacco is still sold in Tianquiz (Fig. C19)

The Aztec, as well as other natives of México, used tobacco for ritual and religious ceremonies, but also for personal enjoyment.[131] Tobacco was carried in the shell of a squash (a calabash) or gourd. Indian physicians or shamans were seen carrying this gourd attached to their waists during religious festivities (Ochpaniztli) and other ritual ceremonies of the god Tezcatlipoca, and during the ritual of puberty. Tobacco was also carried by the priests (*tlazolquacuilli*) to clean and prevent contamination of the temples.[132] They smoked tobacco in pipes, cigars and cigarillos, or inserted tobacco in reeds or *acayetl*.[133]

For tobacco to act as a sacred weed, the Aztec believed that it was necessary to go through a series of rituals. These started at the moment of planting and continued through cultivation. If these rituals were not performed, the tobacco was considered ineffective for religious and medicinal purposes. This transformation could only be achieved by the priest (shaman), would allow the gods to operate through the mystic properties of the sacred weed. There were other plants that were considered sacred because of their psychotropic effects: pulque, which was obtained from the juice of the agave cactus plant, peyote, henben, (hysocimus nigricans) hallucinogenic mushrooms, teonacatl, seeds of the morning glory plant ribea violacea, ribea corymbosa, peregrine, and others.[134]

Tobacco was used during epidemics of smallpox, because of its supposed fumigating properties, smallpox was brought to México by the Spaniards who then perpetuated this habit throughout México, and eventually brought it to Europe where it was used during the plague epidemics of the 17th century.[135]

The Aztec used tobacco in a way unknown to other Indian cultures—they chewed a mixture of tobacco and lime, called *tenex-yetl*. This practice persisted throughout México. It is now known that tobacco releases its active organic acids, including nicotine, when mixed with an alkaline substance such as lime in a process known as "debasing." (Interestingly, the Incas had a similar habit of mixing coca leaves with lime, which releases cocaine.) The tenex-yetl, called *tenchiete* by the Spaniards, was prepared by mixing dried tobacco leaves in a ratio of ten parts of tobacco to one part of lime.[136] This mixture was often placed between the inner cheek and the gum, and it is still used in this way by the Yanomamö Indians of the Amazon.[137]

Fig. 26a. Aztec tlamenetls.

The Aztec tlamemetls (tamemes) were humans trained to carry heavy loads up to 100 to 125 pounds for long distances, 15 miles or more, since the Aztecs and other American people did not have "beasts of burden." The tlamemetls were given tenex-yetl to chew in order to remove the feelings of hunger and fatigue, thus allowing them to make long marches carrying these heavy loads, and it also kept them in good spirits.[138]

King Philip II of Spain sent Francisco Hernández, a botanist, to México to investigate the medicinal and other plants of the New World. Hernández remained in México from 1570 to 1576 and published 16 volumes on his findings. He mentioned that he traveled with Aztec shamans or "*yerberos*" throughout México, and based on this experience he learned that the Aztec used 1,200 medicinal plants, including tobacco. Some of these plants are still used by the descendents of the Aztec.[139] Motecuhzoma had a botanical garden in his vacation resort in Oaxtepec Morelos, close to present-day Cuernavaca, México.

Fray Bernardino de Sahagún (16th century) (Fig. C20) mentions, "whoever sells picietl [tobacco, N. rustica] crushes the leaves, mixes them with lime and rubs this mixture with his hands." Sahagún was one of the first to describe the two tobacco plants—nicotiana rustica (picietl) and nicotiana tabacum (yetl). Of picietl he mentions, "when placed in the mouth [it] produces dizziness and stupefies and on the other hand, yetl was sweet and the smoke was like a ceremonial incense and with healing properties, healed abscesses, sores, colds, snake bites, convulsions and internal disorders."[140] He also mentions that "some are accustomed to take it in small clay or silver pipes or pipes made of hard wood. Others wrap the tobacco in a corn husk in a paper or in a tube of cane [cañutos]."[141]

The Aztecs had pipes made from many different materials: silver, tortoise shell, clay, onyx, stone or reed, with many designs that often represented the particular ritual for which the pipe was used. These pipes were sold in the market of Tlatelolco, which was visited by Cortés and his soldiers soon after their arrival in México City. Some of the ceremonial pipes were beautiful and highly decorated, as can be seen at the Museum of Anthropology in México City. (Fig. 26) (Figs. C21, C22) The archeologist Thomas Joyce wrote in 1914 that after a battle, soldiers of the Maya and Aztecs who had distinguished themselves in battle were given tubular pipes.

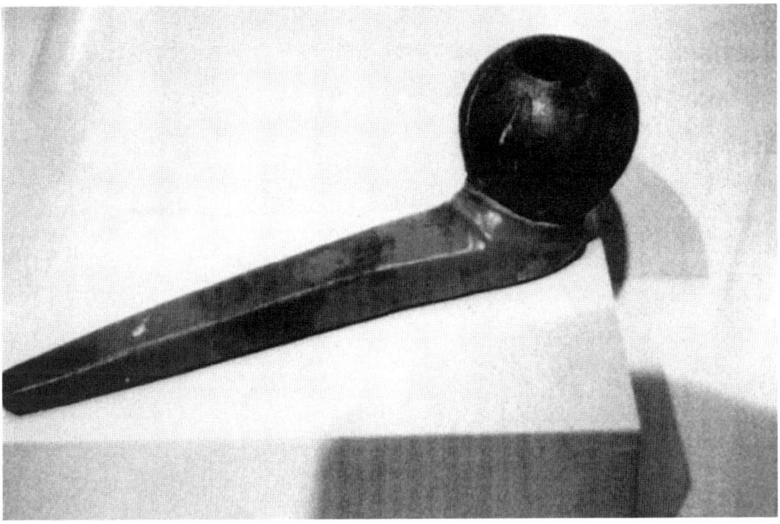

Fig. 26. Mexica angular (stone) Pipe. National Museum of Anthropology, México City. Photography by R. Oropeza.

According to Sahagún,[142] the Aztec priests wore a green jacket (*xicolli*) and carried a ''calabaza'' (calabash) full of picietl during religious ceremonies and sacrificial rituals in the temple of Huitzilopuchtli. The green jackets had skulls and bones printed on it. The calabash was carried on the priests' back, and green balls hung from it. The Tarascan and Huichole Indians of México had a similar custom during their dances and sacrificial rituals. The yetecomatl and the sacred calabash were a symbol of maturity and a sign of respect, a tradition that went way back to the Toltec culture of Teotihuacán. (AD4-8th c)

The Aztec neighbors from Tlaxcala had similar customs, offering tobacco to their god Camaxtli.[143] The Aztecs' captives were made to carry ''perfumadores with yetl [tobacco]'' and ''sahumerios'' (scented urns with tobacco). (Tezozomoc)

Their neighbors of Texcoco smoked tubular pipes after eating and drinking their ''cacao'' (chocolate-tlacacahuatl). They would add ''liquidambar'' to their pipes to sweeten the tobacco. In the house of King Nezahualcoyotl of Texcoco,[144] people brought *acayetls* (tubular pipes) from Tulancingo to smoke and give scent to his home. They used smoke for relaxation, after eating, and during war ceremonies.[145] According to the historian Tezozomoc, the captives of the Aztec brought smoking yetl (tobacco) with them when going to be sacrificed, and the priests smoked their pipes during these sacrificial ceremonies.[146]

"The fragrant cacao has
come to a foam the scented
tobacco has been given
away."
Tlaltecatzin, Aztec poet, 15th century

In a similar manner the Aztec fishermen made offerings to their god Opochtli with foods, wine (pulque), sweet cane of corn, flowers, and

Fig. 27. God Opochtli of The Fisherman wearing on his back a fisherman's net and holding a cigar. Florentine Codex. (Sahagún) 16th C.

tubular pipes with yetl (tobacco). (Fig. 27) The divine spirit was the tobacco spirit.[147] In the mural of *Tepantitla* in the ruins of Teotihuacan (Fig. 28) we can still see priests carrying "offering urns" (*yetecomatl*) of burning tobacco or copalli (copal) as an offering to their gods. To these natives, as in all cultures of North and South America, tobacco and smoke had a divine force to communicate with their gods.[148]

Fig. 28. Tlalocan. Tlalocs' realm. L. Yetecomatl a censer or ceremonial urn. Probably the mother of the pipe. Segment of the Mural of Tepantitla, Teotihuacan, México.

Use of Tobacco in Private, Civil, Religious, and Sacrificial Ceremonies

Fray Bernardino de Sahagún described one Aztec ceremony:

> When a wealthy merchant [Pochteca] wanted to impress other merchants in the community and to give thanks to the gods for all he had received, he would make a large feast or banquet. He would buy all the food and whatever was necessary for the occasion and would invite all his friends and relatives. They would have to be instructed in how to handle the pipes properly during the ceremony, (with the pipe held in the right hand and in the left a special plate or Yiecaxitl on which to rest it). The host would say "my friend here is your perfumed pipe" and they would start to puff. During this ceremony they would offer flowers to the god "Huitzilopuchtli" and they would start to whistle by putting the bent little finger under the tongue and would start burning incense [*copal*]. The guests would be given hallucinogenic mushrooms [*nanacatl*] in the morning, which they mixed with honey. At dawn they would also drink cacao [chocolate] and would start to dance under the hallucinogenic influence of the mushrooms. In the afternoon they would gather and exchange impressions of the hallucinogenic visions they had experienced. The next morning they would bury the ashes of the sacrificed slaves together with the flowers keeping the bone femur and the pipes, they would sing special incantations to the gods, hitting the drum [*atabal*] and they would smoke their pipes and drink.[149] (Fig. C23)

A similar ceremony was done when a son became engaged. The father smoked the pipe when he introduced the son.

In some of these ceremonies human flesh was eaten. Slaves that were called *Tlaltiltin* (*Nahuatl*), meaning "cleaned out" (*Lavados* in Spanish), were washed daily and fattened for a year so that their flesh would be tasty and "clean" when they were eaten. The Tlaltiltin were brought to the market of Atzcaputzalco, now called Atzcapotzalco, after they had been captured during the "flower wars"—wars that were started for the sole purpose of capturing people for sacrificial rituals.

They were dressed properly for the occasion by their owners to attract good prices. They would dance to the tune of singers hired for the occasion. Those who danced with special skill and had "attractive proportions" yielded a higher price, up to forty loads of cotton. The guests at the ceremony would be offered cacao to drink and a variety of carved pipes to smoke. When the slaves were taken to the temple of Huitzilopuchtli to be sacrificed at dawn, they were made drunk from a beverage called *Teuuctli*. The night before they had been given a party, they were fed, and allowed to dance and enjoy music. Before being sacrificed they were dressed in a special garb called *teuxicolli*. They were blindfolded and given only four mouthfuls of a special soporific dough mixed with honey [*Tzoalli*], only four mouthfuls. While the ceremony was going on the priests, host and guests would smoke their carved pipes. One to five slaves would be sacrificed. In the temple they were placed on their backs and held by four men on a special stone [*taxon de piedra*] and the chest opened with a special obsidian knife (*pedernal*). The heart would be removed while still beating and the body pushed down the stairs of the temple. (Fig. C23)

The sacrifice and the removal of the heart had a religious significance. This offer to the gods would give strength to the captor of the sacrificial victim. The captor generally was involved in the ritual. The victim's body would be prepared for the banquet at the home, where it was cooked with corn and salt without chili. This was called the Feast of Panquetzaliztli. As part of the ritual, the clothes and hair from the top of the head of the sacrificed slaves were kept in a chest. During their year in captivity, the slaves were indoctrinated to smoke pipes and learn poetry, dancing and singing.[150] (Fig. C23)

In central México, according to José Péres de Barradas, tobacco was related to the god Tlaloc. The "Curandera," or Herb Woman, in México still invokes the god Tlaloc during these ceremonies. It was believed that when the priest smoked, he brought the rain clouds. According to the historian F. Ortíz, the Aztec thought that there was a connection between the smoke of tobacco and rain, a belief that still exists in some small communities of the descendents of these Indians, who conduct their magic ceremonies by blowing tobacco smoke to the four cardinal points.[151]

According to Sahagún, the dress of the goddess Cihuacoatl ("goddess of the earth, or woman snake") was made of tobacco leaves. This goddess sometimes transformed herself into a snake, and at other times into a beautiful woman who walked in the markets and bewitched young men who were then slaughtered.

Fray Bartolomé de Las Casas wrote that in the festival of Corpus Christi (which is still celebrated by the Indians of Tlaxcala, even though they have been converted to Christianity), there was an abundance of flowers and ceremonial dances. During these festivities they handled large and powerful snakes "as large as an arm and as wide as a wrist with the ease which they would handle a bird. They made these snakes sleepy and dizzy by blowing smoke of tobacco into them."[152]

Medicinal Use of Tobacco by the Aztec

All plants possessed an animated spirit in the beliefs of the Aztec. Plants had a spiritual life of their own. This religion was a form of animism.[153] (anima-soul) Tobacco was an important medicinal plant and was considered sacred by the cultures of central México. The Aztec had a combination of magic, and herbal medicine. They made religious invocations while smoking tobacco and or using hallucinogens.[154]

Tobacco was mixed with other herbs to cure "a swollen head." The medicine man invoked the winds to aid the cure. Blowing four puffs of breath of tobacco smoke on the patient's head concluded the ceremony. Eye pain was treated by "bathing with cold water and an invocation to Chalchiutlicue, the water goddess. The sap of mesquite bark was also applied to the eyes after they had been scratched. The eyelid and eyebrows were then rubbed with tobacco, and the blood from the roots of a cock's feather was rubbed on the eyes. Copal (incense) mixed with tobacco also was used when eyes were scratched.[155] Tobacco enemas were administered with a rubber clyster when a person had general listlessness, pain all over the body, or constipation.[156] The Aztec had many other medicinal uses for tobacco, some of which were transmitted to Europe after the discovery of America. (Badiano Codex; Francisco Hernandez).

During pregnancy women placed tobacco on their bellies as a symbol of good luck. The grandparents organized a feast when the child was born, and smoked tobacco when the umbilical cord was cut.[157] The Aztec also used tobacco as an amulet (or charm) for good luck.

Teopaxtlli was a sacred medicine. It was a concoction made of venomous insects and snake powder mixed with ashes from black pine (*hollín de ocote*), tobacco, and the hallucinogenic *ololiuhqui*. It was prepared as a paste in small sacred vases, and then rubbed all over the body.[158]

Tobacco Miracle Medicine of the Aztec

The Aztecs had all these uses for tobacco: as medicine, to kill snakes, to get "high," to relieve tiredness, to wear as an amulet, to give as offerings, to declare war, and as part of religious and civil ceremonies.[159]
Ailments that tobacco was used to treat:
Toothaches, fever, ear aches, chills, chest pains, worms, headaches, eye infections, snake bites, coughs, insect bites, cysts, rheumatism, birth pains, asthma, convulsions, intestinal ailments (Badiano Codex.)

The following fictitious story is based on an observation of a healing ritual by Leah Brenner. She gives an allegoric description of an herb woman (a curandera) performing a *limpia* (conjuring up evil spirits) on a girl that has been cursed (*mal de ojo*, or "evil's eye"). The girl is a friend of a boy by the name of Pancho, who is watching this "cleansing out" (limpia) ceremony. The illustrations in Brenner's book are by the Mexican painter and muralist Diego Rivera. (Fig. 29)

> The young girl has been suffering from a "jerking eyelid" and her parents take her, with reluctance, to be taken care of by a Curandera (she-shaman). María the herb woman has diagnosed her illness as being a curse. . . . "I can rid your child of the curse, but we must wait until tonight because the moon will help me to cure her with magic." The old man [her father] scratched his head and said, "Go ahead."
>
> The peasants that were present became fearful when María was about to begin the cure. . . . They were fearful that the evil spirits would come into the hut. The old man sitting on the right put his big "sombrero" tilted down to prevent being in touch with the evil spirits, and the mother and the other woman wrapped themselves in shawls and squatted to observe "*la limpia*" [the cleansing or healing]. María has brought the necessary paraphernalia (power tools) for this occasion—her sacred bundle, two eggs, herbs (*jarilla* and tobacco, among others). The healing ceremony is being done on a straw mat (*petate*).
>
> When the moon was full on top of the mountain, she shouted an incarnation to the god Tlaloc [God of the rain that is invoked through the tobacco smoke], "Holy spirit of the god of my ancestors . . . Lord Tlaloc . . . please tell me how to cure this girl with magic." María bends over her sacred bundle and pulls out two eggs, a dead snake and tobacco. She goes into a spell and gets the answer from her god; an evil spirit is making the girl's eyelid jerk. Then María passes the egg over the cursed girl's eyes, murmuring, "May the eggs suck up the evil spirits." [The natives of México, like natives of other cultures of America, believed in transfer medicine.] She cracked the egg and got rid of the bad spirits. [The evil spirits had been transferred from the girl to the egg.] She then rubs the dead snake on the

Fig. 29. Pancho (Diego Rivera himself) observing a 'magical curing ceremony'. From Brenner's story.

girl's eye and blows tobacco smoke in her eyes, saying gently, "My child, your eyelid will jerk no more."

The father looked at his child and screamed, "Her eyes are not jerking anymore! Doña María has rid her of her curse! Then they all went to the back of the hut and enjoyed a jug of *mezcal* (agave juice). The young girl went to hug little Pancho [in the background, obviously Diego Rivera himself], and he happily went to play with her.

Brenner gives a good, clear account of how a "limpia" cleansing is done by shamans in México, who use a combination of magic and herbal medicine during these ceremonies. These types of ceremonies are still common in small communities in rural México.

In the state of Veracruz, in Laguna de Catemaco (Catemaco Lagoon), an annual meeting of witch doctors (*brujos*) and herbal shamans (curanderos) meet to exchange their experiences. The witch doctors practice black and white magic; the herbal doctors practice what has been described above—a combination of ritual religion and herbal medicine.

Cultures of the Occident: Tarasca

The Tarasca

This culture extended to the present states of Michoacán, Jalisco and Colima, México, close to the lakes of Pátzcuaro and Chapala.[160] A good account of the habits of this culture as related to the use of tobacco and pipes was given by the Spaniards during the rule of Viceroy Mendoza and Bishop Vasco de Gama. The Tarasca priests used the pipe often for religious and civil ceremonies, and to a lesser extent for personal use. Tobacco was called *andamuqua*, and sometimes they mixed it with bark from the *amate* tree.[16]

We can still admire the magnificent ruins of Tzintzuntzan close to Lake Pátzcuaro, where they often had their religious rituals. They were a militaristic society, well organized and never conquered by the Aztec.[162] They used copper and bronze extensively and were pottery makers. We can still admire their beautiful copper and wooden crafts and pottery in small communities around Morelia and Guadalajara, and enjoy their unique alcoholic beverage from the tequila cactus (agave).

Tobacco was smoked extensively in pipes, mainly by priests or shamans during religious and civic ceremonies or for medicinal purposes. They made the most beautiful pipes for religious ceremonies, and had ceremonial urns for burning tobacco and/or incense. Beautifully carved pipes of this culture can be seen in the museums of anthropology in México City, Morelia, and Guadalajara.[163] (Fig. C24-25-26) Most of the pipes date to between the 8th and 13th centuries A.D.[164] The Tarasca priests carried their tobacco in a calabash.[165] They wore special clothes that identified them as priests, and they would preside over sacrificial ceremonies smoking their pipes in their temples at Tzintzuntzan ("place of many hummingbirds"). Huitzitila for the Mexica[166] There were no women priests; they were all men but sometimes disguised as women, as in other cultures of Mesoamerica, symbolizing "Mother Earth."

The Tarasca and the Aztec also gave great importance to tobacco as a symbol for declaring war on other tribes.[167] During a war ceremony dedicated to the god of fire, the priests would burn copal and/or the weed *andamuqua* (tobacco) in *sahumerios* (braziers) while "chanting to the gods of morning star, golden morning and god of the red face," and mentioning the names of the cacique enemies. A priest would then make a speech, draw a map on the ground showing the position of the cities that they were going to attack, and explain how they were going to get there.[168] Two spies (scouts) would go and place two balls of tobacco, eagle feathers, and two arrows painted with blood in the city, temple or main house of the enemy tribe. By so doing they would perform "witchcraft" (*hechizo*) and make a declaration of war.[169]

When a Tarasca king died, his main servant would carry his personal belongings, while others carried his perfumes and scented pipes (*cañutos perfumados*). The servants smoked pipes during this occasion. The pipe smokers were called *chacuaqueros*. The chimneys of houses in Guadalajara are now called *chacuacos* to make a symbolic comparison with the pipes of the Tarasca.[170] Soon after the colonization of Michoacán, where the Tarasca culture had flourished, Spanish priests came to convert the heretics to the Christian religion and teach them good morals and manners. Soon, however, the Spanish priests took to pipe smoking on their chacuacos (pipes).

Cultures of the South (West)

Mixtec culture—Some pipes had a symbolic meaning as we can see in the pipe from a mourner. The bowl of the pipe is represented by a skull. (Fig. C27)

The Mixtec used, and still use extensively hallucionogenic-psychodelic mushrooms in their religious ceremonies. María Sabína attracted shamons, scientists and the common folk, on her 'mushroom veladas' that had christian overtones. Peyote was, and is still used in their ceremonies.

6.

CULTURES OF SOUTH AMERICA Amazon, Orinoco, North Coast, Inca and Gran Chaco (Argentina)

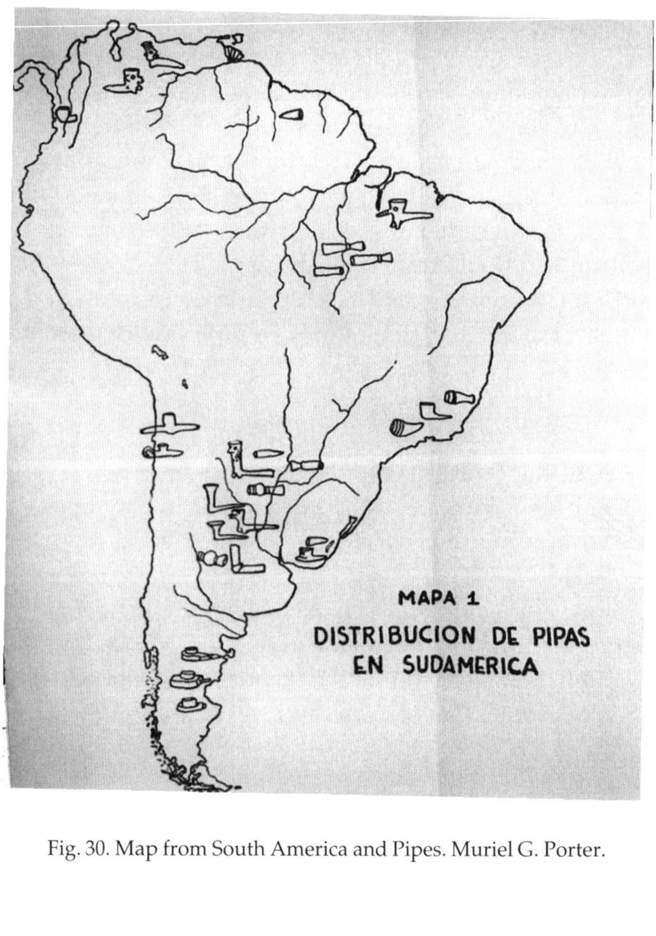

Fig. 30. Map from South America and Pipes. Muriel G. Porter.

The People of the Amazon and Orinoco River (Brazil-Venezuela)

They smoked tobacco in pipes, chewed it, and inhaled it with other hallucinogenic powders such as henben or A.peregrina.[171] (Fig. 31) The same

Fig. 31. Carib inhaling tobacco smoke and other stimulant or hallucinogenic drug from a gourd. Museo del Hombre Dominicano. D.R. Photography by R. Oropeza. 1993

habits were practiced throughout the north coast of South America (Venezuela, Margarita Island, Colombia). The largest concentration of pipes has been found in present-day Argentina in the area of Gran Chaco.[172] (See Map)

Tobacco was so scarce that they mixed it with the bark of trees, like North American Indians. The Inca west of the Andes knew tobacco by the name *Sairi.* They inhaled it only for "medicinal purposes," but there is some evidence that they also mixed it with hallucinogens or stimulants (like *coca*). They depended on coca to obtain similar or "better" effects than tobacco, and that is probably why they did not use tobacco extensively.[173] They smoked tubular and angular pipes, a few made of gold.

They have been found by Hamy in Antioquia, Colombia. Most pipes were made of clay.[174] In present-day Brazil they smoked tobacco mainly as cigars, but they also chewed it and smoked it in pipes. On the southern part of the Andes (Chile), they smoked tobacco in pipes.[175]

André Thevet (1509–1590), a Calvinist French Huguenot, went with 600 Calvinist Huguenots in 1556–1557 to settle in the present-day country of Brazil to escape religious persecution in France. He and many of his followers became smokers. The Huguenot colony failed to take root and came into conflict with the Portuguese settlers. Thevet returned to France in 1558, bringing N. Tabacum tobacco plants and seeds to his native town of Angoulême. (The plant first became known in France by that name). Thevet was not given recognition for introducing tobacco to France; that "honor" was given instead to Nicot. Thevet wrote: "The Christians that now inhabit these lands have become very desirous of this herb and its perfume, although the first use of tobacco at the beginning is not without danger before one is accustomed to it, for this smoke causeth sweats and weakness even to fall into a swoon" (as happened to him.)[176] He mentioned that in spite of the undesired effects of smoking, they kept coming back for more. Obviously, they were already addicted.

Amerigo Vespucci was a contemporary and acquaintance of Christopher Columbus. Vespucci was an educated Italian merchant and navigator. He determined that the lands that had been found by Columbus were *mundo novo* (new lands), and not the land of Cathay as Columbus thought. Vespucci drafted maps while exploring South America under Portuguese patronage in 1499. He later became a Spanish citizen and navigator in 1508.[177] The name "America" was given to the new continent in recognition of his identifying these lands as "new."

In 1499 Vespucci entered the Amazon River. Later, he explored the Orinoco River and the north coast of South America, reaching the Margarita Islands (Venezuela) before Sir Walter Raleigh.[178] There he made his observations of the natives chewing tobacco. On his second expedition (1501) he followed the coast of Brazil and the Río de La Plata (Argentina).[179]

Upon observing the natives of Margarita Island chewing (sucking) tobacco, Vespucci reported:

> In behavior and looks they were repulsive, and they all had their cheeks swollen with a green herb inside, which they were constantly chewing like beasts so that they could scarcely utter speech. They carried two dried gourds hanging from their necks, one with the herb, the other with a white

flour, which looked like powder chalk and from time to time with a small stick which they kept moistened in their mouths, they dipped it into the flour and then put it into their mouths inside both cheeks, thus mixing with flour and herb.[180]

One can assume that the "powdered chalk" was either lime or a powder obtained from sea salts or cockleshells mixed with potato paste or ashes of the acacia tree. Both of these substances, which are alkaline, increase salivation, which is also alkaline, and thus help to release the nicotine from the tobacco. Vespucci also mentions that only men had this habit, and assumed wrongly that they were doing this to quench their thirst. Some of these natives (the Yanomamö) still put a wad of green tobacco inside their cheeks.

Indians of Present-day Brazil (Tupinamba)

Pedro Álvares Cabral, born in Belmonte, Portugal, was named one of the foremost Portuguese navigators by King Emmanuel I of Portugal. Cabral was given the order to follow the conquest of lands earlier acquired by Vasco de Gama. He was named admiral of thirteen ships, which set sail from Lisbon, Portugual, in 1500. On April 22, 1500, he accidentally landed in what they first named Holy Cross, but which they renamed Brazil because of the wood of that name (*pau brasil, palo brasil*) found in this land.

When Cabral explored the coast of Brazil, he noticed pipe smoking by these Indians. (Fig. 32) We may assume that he and his sailors took with them the habit of pipe smoking back to Portugal. Nicholas Barré in his description of this trip in 1555 mentions that Cabral said, "I have seen a plant that they call *petun* or *petum*, the size of a large confrey, (yellow henbane-beleño, hyoscimus nigricans.) They suck the juice and inhale the smoke of this. With this plant they can endure hunger eight or nine days."[181]

Ferdinand Magellan, under a Spanish flag, sailed by Brazil and the tip of South America (Patagonia) in 1519 on his circumnavigation around the world. He carried tobacco with him to the port of Cebu in the Philippines, where he was killed in Mactan. This was the first introduction of tobacco to Asia.[182]

Damiâo de Goes (1501–1574) an educated man in the Court of Portugal under King Emmanuel, exchanged ideas with the intellectuals of his time such as Erasmus from Holland, and Luther and Melanchton from Germany. De Goes gave the first tobacco plant and seeds to Jean Nicot, ambassador from France to Portugal, who became familiar with the use

Fig. 32 Tupinamba of Brazil during a Petún (tobacco) dance ceremony in 1621.

of tobacco. Tobacco was held to have a curative value in the treatment of cancer and other illnesses, and Nicot became a promoter of tobacco for medicinal and other purposes in France. He prescribed it to cure the headaches of the queen mother, Catherine de' Medici. This was the beginning of the use of tobacco by the royalty and aristocracy of France. It became "chic" to snort tobacco because the queen had adopted the habit and encouraged this "barbarian" custom of the Indians of America.

De Goes mentioned that Cabral described tobacco on his return from Brazil as "an odoriferous and medicinal plant that the natives of Santa Cruz (Brazil) called *fumo*." They also called it *petun* or *betom*. Cabral named it "the holy herb" because of its many supposed virtues that he said could "cure abscesses, sores, fistuli, polyps and other ailments."[183]

Cabral (1500) also mentioned some of the habits of other natives of Brazil—the Tupinamba. Cabral observed that these people carried a calabash made in the shape of a man's head with a mouth, nostrils, eyes and hair. "[They] placed on the top of an arrow [a pipe with a long stem] with which they made smoke with dried leaves of the plant petun, and the smoke which is in the head they inhale to such an extent that they get drunk." Cabral mentioned Indian rituals in which they smoked and inhaled tobacco, and he stated that the effects on the participants were violent. We have to assume that these natives mixed tobacco with the hallucinogen *piptadenia*, or another hallucinogenic substance like henbane,[184] which was and still is customary among many of the indians of South America. (Yanomamö)

One of the present-day tribes descendent from those observed by the first explorers is the Yanomamö, the so-called "Fierce People" of the Amazon. Their population numbers about 10,000. They still practice the habit of sucking tobacco by placing wads of tobacco inside their cheeks.[185] (Fig. 33) They also inhale the hallucinogen *eben* (henbane), that is obtained by scraping the inner bark of that plant. The substance is then moistened by having the women chew it and spit it out (saliva is an alkaline substance). Then the paste is mixed with wood ashes. By mixing the ashes of certain trees, such as acacia, with saliva (both substances being alkaline), the active hallucinogenic substance from the bark of the eben (henbane) plant is released. The substance is then dried and ground into a powder. This powder is inhaled into the nostrils. Individuals sit or squat opposite one another inhaling the substance until they have "color visions" (psychedelic hallucinations). According to Chagnon, men, women and children all chew tobacco and probably mix it with piptadenia or henbane.[186] (beleño)

'These people cultivate tobacco and jealously guard their parcels.' Stealing of tobacco is very common, so fences are built between owners—even between fathers and sons—so that they cannot go into each other's property without permission. Sharp fragments of bone are placed in the ground to prevent trespassing.[187]

Sir Walter Raleigh was one of the confidantes of Queen Elizabeth of England. She was very attached to him and he was probably her lover. When Raleigh got married he had to do it in utmost secrecy. When his son was born, the queen found out about it and became very jealous and upset.

Raleigh was an educated man, a consummate soldier who helped the religiously persecuted Protestant Huguenots in France, an admiral of the seas, and an explorer. In 1595, under Queen Elizabeth, he made an expedition to present-day Guiana, South America, sailing up the Orinoco River[188] looking for "El Dorado."[189]

Raleigh was very smart but often caused intrigues in the court. After Queen Elizabeth's reign ended, (1603) Raleigh got into trouble when he defied King James I. In 1617–1618 Raleigh made a second expedition that led to his death by guillotine. He was looking for gold mines, but he failed. While doing so he was supposed to avoid entering Portuguese or Spanish colonies. One of his lieutenants, Lawrence Kemys, burned a

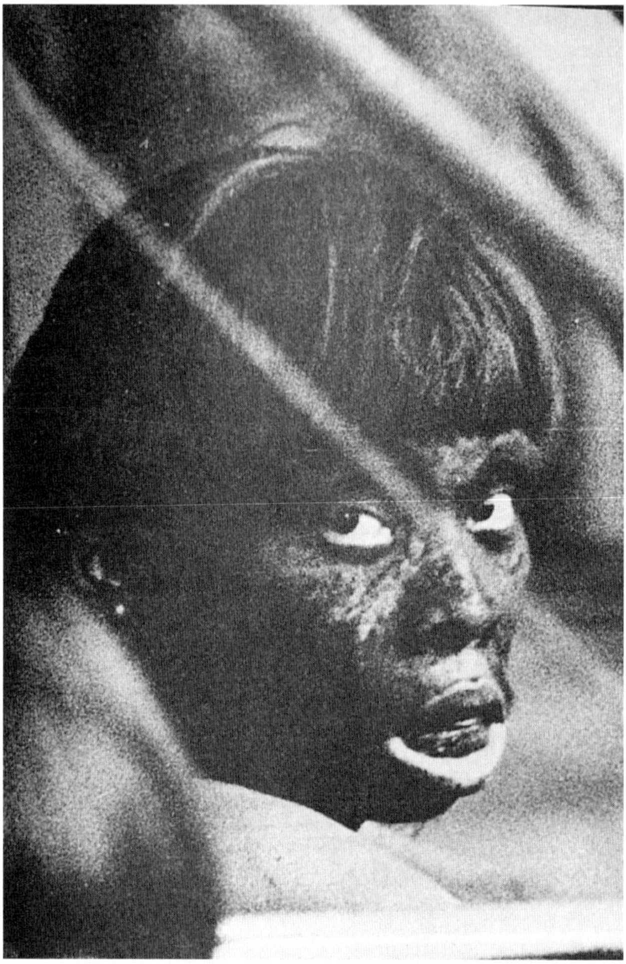

Fig. 33. Yanomamö looking fiercely through the string of his hunting bow, a wad of green tobacco protruding from his mouth. This practice was observed five centuries ago by the Italian explorer, Amerigo Vespucci. From Napoleon Chagnon's book, The Yanomamö.

Spanish settlement, creating a diplomatic conflict with Spain.[190] Raleigh had been imprisoned before in the Tower of London, but was given a pardon in 1603 by King James I. This pardon was rescinded because of the incident at the Spanish settlement. Raleigh went defiantly to the guillotine proudly smoking his pipe. (John Aubrey) This was his way of getting even with King James I, who abhorred smoking.

During his first Guiana expedition of the Orinoco River up to St. Thomé[191] (present-day Bolivar, Venezuela), Raleigh observed the natives chewing and inhaling tobacco. He wrote that they were "tribes handsome and healthy who were greatly addicted to tobacco."[192]).

One of the dreams of this author after reading Raleigh's Second Expedition, was to follow his path. When the opportunity came in the summer of 1970, this expedition was planned with a group of physicians from Venezuela headed by Dr. Francisco Romero and a native of the upper Orinoco who had been removed from his tribe and adopted with the consent of his parents when he was only a child and later educated to become an engineer.

The trip was done by canoeing on and off thru the tributaries of the upper Orinoco River to reach the Caroní Falls, passing close to the Angel Falls and finally reaching the Canaima Falls in the Venezuelan Guiana.

We observed the customs of these wavao and other tribes, the use of medicinal plants and we observed how they suck—young and old, men and women—a wad of green tobacco. They put their tobacco wad under their lower lip or inner cheek giving a disgusting look, as had been described 500 years before by Cabral, Amerigo Vespucci and others. (Fig. C28)

Emanuel Van Meeteren (1536–1612), a Dutchman, also observed these customs. He mentioned how in Brazil they smoked "pipes especially made of baked clay one finger long and in front a little bowl about the size of a hazelnut which they fill with the powder and then light it with a live coal or candle and at the other end they suck it in a hollow pipe smoke."[193]

Jean De Léry in 1557 described a priest (shaman) of the Tupinamba tribe as "often taking a wooden cane four or five feet long, at the end of which there is some of the plant petun, dried and lighted, and turning around and blowing the smoke to others." De Léry stated that they "received the spirit of power, that you may conquer your enemies." De Léry tried tobacco and said that "he felt refreshed and [it] kept him from feeling hungry."[194]

The explorer Girolano Benzoni, during his 1565 exploration of the north coast of South and Central America (present-day Venezuela, Colombia, Costa Rica and Nicaragua), described the use of tobacco by the slaves from Africa who inserted tobacco in a rolled husk of corn, set fire to it and smoked it. He said that some of them fell to the ground and remained unconscious most of the day after smoking.[195] (See Fig. 9)

The Inca (1100–1531)

Francisco Pizarro (1531) and the Inca Civilization of the Andes (Peru-Bolivia)

A Psalm to His Own Death (Pachacuti's)

"I was born like a lily in the garden
And so also was I brought up
As my age came, I have grown up
And, as I had to die, so I dried up
And I die."
Pachacuti (1470, Inca), Perú

"Each leaf is fruit and such substantial fare
No fruit beside to rival it will dare."
(Dr. Abraham Cowley)

Soon after the conquest of the Inca Empire, Garcilasso " 'Inca" de La Vega (16th century), a descendent of a relative of the last Inca king, was educated in Spain and became an eminent authority on the Inca culture. His father was proud of his ancestral Spanish origin, his experience in battle, and his intellect in literature. He was in Pizarro's army, and after the fall of the Inca Empire the women were divided among the conquerors. He received the niece of a prominent Inca, Huayna Capac, cousin of the famous Inca Emperor, Atahualpa. The children of noble ascendancy were sent to Spain to become educated, and in this way the Spaniards got firsthand information on the customs of the Inca. De La Vega's writings are considered accurate, informative and invaluable.[196] He did most of his writing while in Spain, where he lived part of his life.

De la Vega and other historians at the time of Pizarro have given us information about the Inca culture and its use of tobacco and coca.[197] De la Vega also tells us about the Inca Emperor's reaction to the conquest:

> Your emperor may be a great prince, I do not doubt it, seeing that he has sent his subjects for far across the waters, and I am willing to treat him as my brother. As for the pope of whom you speak, he must be mad to speak of giving away countries that do not belong to him. As for my faith, I will not change it. Your own god, as you tell me, was put to death by the very men he created. But my god still looks down upon his children.
> (Atahualpa, the Inca ruler, in response to Pizarro telling him that Pope Alexander VI had declared Perú to be the possession of Spain, 1533.)[198]

De la Vega described the Inca market as follows:

> The Inca market like the Aztec market had many luxury items. The trade was a government monopoly unlike the Mexicans' which was a private enterprise. From the coast came cotton of different kinds that was easy to weave and there was a number of dyeing colors. *Achiote* to give a carmine color to the seashells, seaweed, seashell food, dried fish of all kinds, reeds to make baskets and many tropical foods. The jungle people brought feathers of all color for their weaving, *chonta* palm, [iron hard wood] birds and many other animals, medicinal herbs, quinine, *ipeca*, sassafras, *quayosa* [rubber], gum latex from *papodilla* tree [*chicle*], strong tobacco leaf [n.rustica] and narcotics [coca and others]. Since there was no coinage, they bartered.[199]

According to Baron von Humboldt, (1799–1804) the Inca knew both tobacco plants—nicotiana tabacum and nicotiana rustica and other nicotiana.

De la Vega (as quoted by Prescott) mentioned that the Inca medical practitioner (shaman) rarely smoked tobacco (*sagri*), but used it as a snuff to clear the nasal passages.[200] The shaman was held accountable if the medicine failed. He was "trussed up together with the patient" and exposed to death with him."[201] More recently, however, Spinden mentions that the Incas most probably mixed tobacco with narcotics (coca) and used it as a snuff for pleasure, as did their Aztec counterparts,[202] or administered this mixture in the form of an enema with a rubber clyster. (See Fig. C4) According to Dr. W. G. Mortimer they mixed tobacco with coca and chewed it, a habit that is still practiced at the present time.[203] So it might have been a perception by the Spaniards, rather than a fact. One reason why the Incas did not use tobacco in great quantities might have been that they found a better substitute in the stimulating qualities of the coca plant (*erythroxcyclum peruvianam*), or *cuca* as they called it.

Mortimer mentioned that coca was also used by the *curadora*, or "healer." She was "an old women who understands the secrets and removes the evil spirits of sickness by means of poltrice of tobacco and mustard together with coca leaves and using certain particular signs."[204] Mortimer, who was a physician, mentioned that the constant chewing of coca caused a swelling in the side of the cheek that they called *picho*.[205] This probably was a cancer of the inner cheek similar to the cancer produced by chewing tobacco with betel nut in India. The curadora (curandera in México) used tobacco in her magical medicine (witchcraft) to conjure up bad spirits. A similar custom was practiced by the Taíno of Hispaniola. The Incas used pellets of tobacco before setting fractures.[206]

Benzoni, while in Perú (1565–1579), gave us his observations on the Inca's use of tobacco combined with coca: "When they will make themselves to be out of judgment, at which time they totter and go as though they were out of their wits, or if they were drunk which is a thing that doth give them great contentment to be in that sorte."[207]

Tobacco, coca and other stimulants were used in the Americas. These practices were passed on to the conquerors. There are similarities between the use of coca by the Inca and the use of tobacco by the Aztec. Both cultures used the leaves of these plants as stimulants to energize them during long walks and arduous tasks such as carrying heavy burdens. Both cultures mixed the dry leaves of these plants with lime or other alkaline substances such as shell powder or the ashes of the acacia tree, which released the active drug (alkaloid) of tobacco and coca. Both measured distances by how many chewings they made either of tobacco or coca. Both cultures used these plants as stimulants in religious ceremonial practices, and both were highly dependent on these addictive plants.

The Inca called coca "Mamma plant" or "Mamma coca." It was the divine plant of the Inca. The two plants—tobacco and coca—were given similar names by the Spaniards in the 16th century when they first encountered the Indigenous people of the Americas. Tobacco was called *herba sancta* (Latin for "sacred herb") and coca was called *foha sagrada* (Old Spanish for "sacred leaf").[208] The two plants were considered to be sacred or divine in all the Indian cultures.

Coca was brought to Spain, probably by the soldiers of Francisco Pizarro. Its pleasurable and addictive qualities were not immediately recognized, however. It was not until the mid-19th century that Angelo Mariani, a Corsican,[209] began experimenting with the coca of the Inca (which was soon to be considered as the elixir of life, rendering the "divine herb" popular.[210] Soon after, artists and composers including Massenet, Fauré, and Gounod, the actress Sarah Bernhardt, physicians including Sigmund Freud and William Halstead, and even Pope Leo XIII became addicted to coca and sang their hosannas to this plant. (Pope Leo XIII was suspected to have taken the coca elixir prepared by Mariani during his "ascetic retirement!")[211] In 1884 Dr. Carl Koller used cocaine as an anesthetic for the first time in eye surgery.[212]

Peoples of the East Coast of South America (Argentina, East of the Andes). The Chaco Culture.

The people east of the Andes did not use coca as the Incas did on the west side of the Andes; however, they did use tobacco extensively. The largest collection of pipes in South America has been found in the area of Chaco (the Calchaqui region). (See map) These Indians—the Tehuelche from Argentina—also chewed tobacco and mixed it with wood fragments, probably to conserve the tobacco, which was scarce.[213] Other natives used tobacco as a snuff.[214] Tobacco was hard to come by in Tierra del Fuego, but tobacco use had no boundaries; it was used universally from the top of North America down to the tip of South America.

John Byron in 1764-1766, grandfather of the famous poet John Byron, led the English expedition searching for the Falkland Islands on his way to circumnavigate the world aboard the Dolphin. Approaching the shores of Patagonia, he decided to encounter these amerindians (patagones or big feet). Scores of these men and women were on horseback. They gathered at the beach where John Byron distributed trinkets such as beads and green ribbons. They appreciated these gifts but they made it clear to him, by signs, that they would rather have tobacco. So Byron sent for some and then the encounter was over. (H. Wallis, The Patagonian Giants, cited in Gallagher, pp. 185, Dickinson, The Circumnavigators, 2003.)

7.

CULTURES OF NORTH AMERICA

A Withered Old Man and a Stalwart Youth
"I blow my breath" said the old man
"And the steam stands still."
"I breathe" said the young man
"And the flowers spring up over the plain!"

At sunrise the youth, who was spring,
recognized his companion as winter, and
as he did so the old man melted away into
nothing, leaving behind him a single white
flower.

Mark Twain, from an American Indian legend referring to the smoking of tobacco.

The people of North America were scattered throughout the vast continent of present-day United States and Canada. They were mostly isolated from the more organized societies of Mesoamerica (the Aztec, Maya and Tarasca).

For the purpose of simplification, the large number of North American tribes have been grouped according to their geographical locations and, to a certain extent, by their cultural similarities.[215] (Fig. 34) Many of these tribes had their own customs and rituals. They all had in common the use of tobacco for many purposes. However these people were heavy pipe smokers.[216] They did not know nicotiana tabacum; they smoked nicotiana rustica and other strong nicotiana. A few tribes in the Southwest smoked or chewed nicotiana attenuata, a plant not cultivated, but which grew wild.[217]

C1. Wild nicotiana growing in the Sierra Madre of Zacatlán Puebla, México. Photography by R. Oropeza.

C2. Nicotiana Tabacum in bloom. Drawing by R. Oropeza.

C3. Ornate Nicotiana. N. Sylvestris or N. Alata Glandiflora. Photography by R. Oropeza.

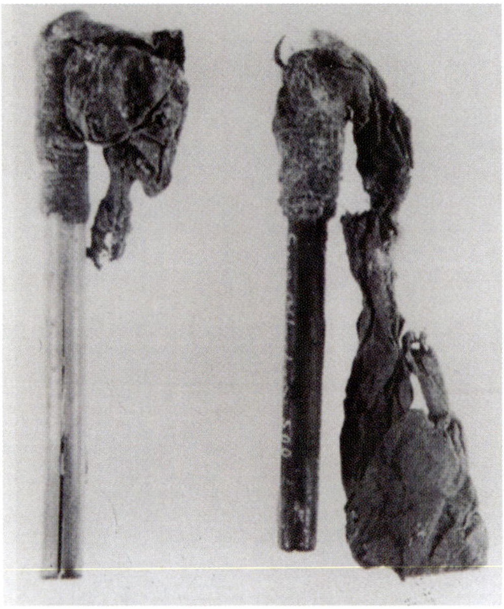

C4. (above) Precolumbian rubber enema syringe from Olleache Perú, S.A. Ethnographic Museum of Stockholm, Sweden. After Wassen.

C5. (right) Huichol tobacco ceremonial pouch. Collection R. Oropeza.

C6. La Rábida Monastery. Palos, Huelva, Spain. Columbus Portico to the New World. Photography by R. Hagan.

C7. (above) The Santa María, one of the three ships that navigated with Christopher Columbus. Palos, Huelva, Spain. Photography by R. Hagan.

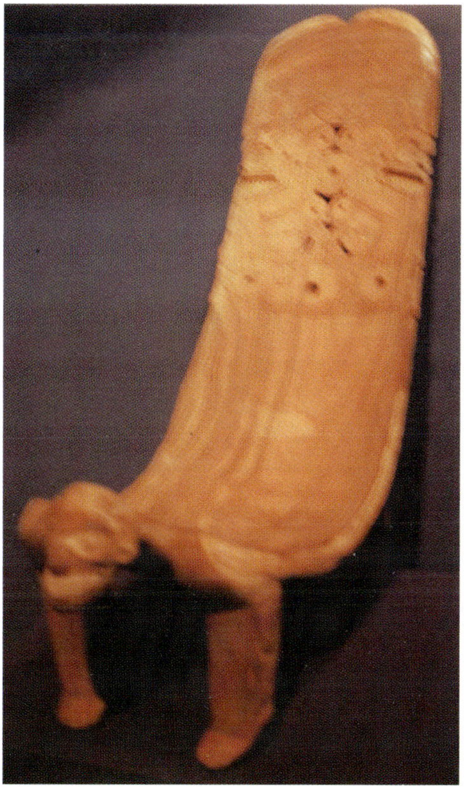

C8. (right) A Duho. Taino ceremonial chair carved from guayacán wood from Haiti. Museo del Hombre Dominicano D.R. Photography by R. Oropeza.

C9. Map. Cultures of Mesoamerica. Museo of Anthropology, Jalapa Vera Cruz, México. Photography by R. Oropeza.

C10. The author beside the head of a giant Olmec head (40 tons) Museo de La Venta, Villa Hermosa, Tabasco, México. Photography by R. Oropeza with a timer.

C11. Map. Cultures of the Gulf. Huaztec-Totonac. Museum of Anthropology, Jalapa Vera Cruz, México. Photography by R. Oropeza.

C12. Totonac Shaman (censer) chewing a roll of tobacco (4th–8th c. AD). He wears a Shaman's headdress. The eyeglasses of Tlaloc, god of rain. He chews on a roll of tobacco, a common practice of cultures of the south. A tobacco pouch hangs from his neck. Museum Rufino Tamayo, Oaxaca, Oaxaca, México. Photography by R. Oropeza.

C13. Totonac warrior smoking a cigar (4th–8th c. AD) probably the earliest artifact of cigar smoking. He wears flower-shaped earflares and a jade necklace, probably a noble warrior. Private collection Rufino Tamayo. Oaxaca, México. Photography by R. Oropeza.

C14. (left) God L. Temple of the Cross. Palenque Chiapas, México. Photography by R. Oropeza.

C15. (above) A recent artifact of the descending god of Tulum. México. Photography and collection of R. Oropeza.

C16. (left) The Reynolds vase probably depicting the legendary Maya twins Hun Hunahpu and Vuqub Hunahpu. He is smoking a slim cigar, sitting in the Lord's (lotus) position. Museum of American Indian, Smithsonian Museum. Washington.

C17. (above) The plate of the smoking husband. Smoking a slim cigar while the wife prepares masa (dough) in a metate, to make tamales. University of Oklahoma Press.

C18. A sacrificial rite and tobacco. (center) Stone of the sun, the Aztec calendar. Sacrifices were carried out on this stone, but more often on the Stone of Tizoc. Museum of Anthropology of México City. (clockwise from top) (a) Purification of the temple and victim before the sacrifice. Self sacrifice was as important to them as the victim's sacrifice itself (Penance). (b) Gladiator tied to the Tizoc's Stone fighting for his freedom. (c) The victim's chest is opened and the heart is offered to the gods. (from Tezozomoc, 1594) (d) The scalp, femur bone and jacket are saved for the ceremony. (e, f) Tobacco is smoked by the main participants. (From Sahagún, 1529–1590). Montage by R. Oropeza.

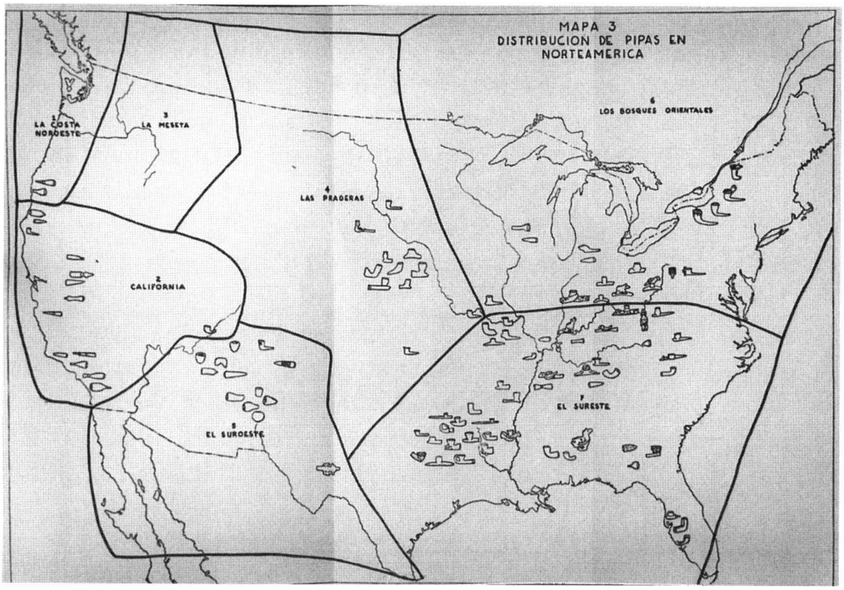

Fig. 34. Map. Cultures of North America and pipes. Muriel G. Porter.

The nicotiana rustica that they smoked was high in alkaloid (nicotine), containing up to 10 percent nicotine in the leaf. It was bitter. The leaves were small and not suitable for cigars, but they could be smoked in pipes. This is most likely why the aborigines of North America developed the habit of pipe smoking, unlike the Maya who used nicotiana tabacum, which is milder, has larger leaves and could be rolled into cigars.

Nicotiana tabacum was brought to the north part of the continent in 1610 by the Englishman John Rolfe (husband of Pocahontas) from Trinidad. (B. Laufer p. 105)

Until the middle of the 16th century, Spain and Portugal had the monopoly on all the goods obtained from the new world, including tobacco, which at that time was already in great demand in Europe. This monopoly was maintained because of the unchallenged power of the Portuguese and Spanish Armadas. In the early 16th century the English had not participated in the exploration of the new world because they had to abide by the Treaty of Tordesillas, since England was then still under Papal rule.[218] At the end of the 15th century, Pope Alexander VI had issued the treaty of Tordesillas (on June 7, 1494), which granted Spain and Portugal the right to the new lands that were or would be discovered in the Americas. This treaty allowed the Portuguese to explore Brazil and take possession of its lands.[219] When England became Protestant in the mid-16th century, it no longer had to abide by this treaty. Sir Walter Raleigh and John Hawkins, under Queen Elizabeth, decided to build a strong navy and challenge the Spanish Armada. Hawkins started a war of piracy in 1562, which ended twenty-five years later with the destruction of this Armada (1588). Hawkins raided the west coast of Africa for slaves that were brought to the English West Indies (Jamaica).

Before the English started to explore and settle the east coast of North America, other explorers including Ponce de Leon, De Soto, Coronado, Verrazano, Cartier, and the French Huguenots had ventured into these lands, but had not established permanent settlements. (We are naming only those explorers who made contact with the people of North America and made reference to the use of tobacco. Other explorers, such as Cabot and Frobisher, never mentioned tobacco in their diaries.)

Tobacco's Mythical Origins in North American Cultures

North American indigenous myths show that all of their cultures believed in the supernatural power of tobacco and its supernatural origins, as illustrated by this Winnebago myth:

> However, Earthmaker did create a weed and put it in our charge, and he told us that none of the spirits he had created would have the power to take this away from us without giving us something in exchange. Thus said Earthmaker. Even he, Earthmaker, would not have the power of taking this from us without giving us something in return. He told us if we offered him a pipeful of tobacco, if we poured this for him, he would grant us whatever we asked of him. Now all spirits come to long for this tobacco as intensely as they longed for anything in creation, and for that reason, if at any time we make our cry to the spirits with tobacco, they will take pity on us and bestow on us the blessings of which Earthmaker placed them in charge. Indeed so it shall be, for thus Earthmaker created it.

The First Explorers of the Northeast Coast of North America

King François I of France (1494–1547) spent a good part of his time dallying in love affairs, but he did not neglect the duties of the state. He was an intellectual who attracted to his court many Italians of importance, such as Leonardo da Vinci, Cellini, Soloni and others, to enrich the arts during his monarchy. At the same time he filled many positions of importance in his government with the best Florentines and Milanese, and his naval force with Genovese and Portuguese navigators. He chose two brothers, Giovanni and Girolamo da Verrazano, for a North American expedition. They were experienced navigators and cartographers.[220] (1523)

The southeast coast of the Americas from Florida to Patagonia had already been explored by the Spanish and Portuguese, but an expected passage to Asia through the north had not been found. Europeans were still looking for a way to establish commerce with Asia (the land of Cathay and Zipango) through an easier route than the one found by F. Magellan (1519) through the southern tip of South America. This was the main impetus for the exploration of the north—to obtain silk, spices and gold. They never expected to find "brown gold" (tobacco) instead.

Verrazano's voyage (1524–1528) was financed by wealthy merchants of Lyon, Rouen, and Dieppe, from where they departed.[221] Landfall was in present-day North Carolina on March 1, 1524. Verrazano's second-in-command was his brother, the mapmaker Girolamo, who called the point of landing "Dieppa."[222] While looking to anchor, they were well received by a group of natives and made the following observations: "These people go altogether naked except that they covered their private parts with certain skins of beasts like unto [that look like] martens which they fasten into a narrow girdle made of grass, very artificially [artfully] wrought hanged about with tails of diverse beasts, which round about their bodies hang dangling down to their knees."[223] Verrazano made a full description of the land he discovered and named it Arcadia (inspired by the ancient land of the Peloponnesus).[224] Eliot Morison has accurately located this area at present-day Kitty Hawk, North Carolina.[225] Verrazano went ashore with some of his men to explore, and he wrote the following: "In Arcadia we found a man who came to the shore to see what [manner of men] we were. . . . He was handsome, naked, his hair fastened in a knot, and of an olive color." Approaching a group of twenty Frenchmen, the native thrust toward them "a burning stick, as if to offer us fire. . . . this was a lighted tobacco pipe. He was trying to make a peaceful overture." His gesture was misinterpreted by Verrazano's crew, however, who believed that the natives were ready to attack and fired a shot from a musket that "made him tremble with fright."[226]

Verrazano continued his voyage northward until he entered the narrows that now carry his name at the present-day Verrazano Bridge, and then continued along the coast of present-day New England until he reached Narragansett Bay. Verrazano stayed for fifteen days in Narragansett Bay and made the following observation of the customs of these people (the Wampanoag): "If they are afflicted with a wound, they heal themselves with fire [tobacco smoke] without outcry."[227]

Jacques Cartier (1534–1535) and the People of the St. Lawrence River, Canada

"It is done, we are brothers.
The conclusion has been reached,
now we are all relatives
Iroquoi, Hurons, Algonquins, and French
we are now but one and the same people."
Jean-Baptist Atironta, a Huron captain, at the conclusion of a 1645 peace conference.
(Alvin M. Josephy Jr., *The Native Americans*)

Jacques Cartier, an aristocratic and educated explorer, a master of the sea, explored where Verrazano had left off and continued farther northward up to the St. Lawrence River in present-day Canada, still looking for a passage to the land of Cathay (China). During his first trip in 1534, Cartier did not mention the use of tobacco by the Huron, but on his second trip in 1535–1536, when he entered the St. Lawrence River, he met some Indians who were returning from distant trading sites. He described these Indians as "savages" or "*peau rouges*" (red skins), and never referred to them as Indians. He mentions that "they carried a pouch with dried leaves of a plant that they called kiyekta,[228] they crumbled these leaves and smoked them in a clay pipe with a reed stem, they lit it with a live coal, and alternately 'sucked' in and blew out great clouds of smoke, like smoke from a chimney. . . . they say it keeps them warm and in good health, and never go abroad without it." The French tried a few drags from a lighted pipe, but found the smoke too hot to bear, "like powder pepper."[229] The explorers were surprised by the practice of smoking tobacco, which is an indication that smoking was still not very popular in Europe at this time.

The Huron were called "the tobacco people." They used tobacco extensively and also used it to barter for other commodities.[230]

During that winter most of Cartier's crew became sick. They were losing their teeth, and had infected and swollen gums and painful legs. They were probably suffering from scurvy due to a lack of vitamin C. The Hurons were familiar with this disease, which occurred during the winter when they could not consume fresh produce. They knew how to prevent this ailment by boiling the bark of a tree called *annedda* (arbor vitae), taking the juice and making a concoction.[231]

Alvar Núñez Cabeza de Vaca and his three companions. The American South from Florida to Texas and New Mexico Taken from the chrnoicles of the Nanváez Expediton by Fannie Bandeliere, 1909

In 1527 Cabeza de Vaca, a Spaniard, left Cuba with Pánfilo de Narvaez and 600 men to explore the coast of the Gulf of México up to Florida. Their boats were shipwrecked (1528) on the coast near present-day Galveston during a storm. Only 60 men survived, and most of them were killed by Indians. After one year Cabeza de Vaca and only three others remained alive. Cabeza de Vaca stolidly undertook the painful exploration of the Southwest for a period of eight years, until 1536. He did reach present-day Arizona and New Mexico and gave an account of the customs of many of the tribes he encountered in a book entitled *Naufragios* (*A Shipwreck*) in 1542.[232] (Popularly know as La Relación)

Cabeza de Vaca described the Malhado Indians of Florida: "Throughout this country they intoxicate themselves with a smoke (tobacco smoke) and they give whatever possessions for it."[233]

The journey exposed him to landscapes previously never imagined by Europeans and he was exposed to many new experiences and observed strange customs of these tribes.

He mentioned in 1539 that people of the present-day Texas used mezcal beans as an article of trade. The tribes he met cultivated corn, beans, squash, and cotton. In Texas they smoked desert-wild-tobacco, N. obtusifolia, sometimes mixed with other leaves.

While passing through the present city of San Angelo, he described the Mezquite tree, the Texans hare (Lupus Texianus), it was called the jack ass rabbit during the Texas war with Mexico. He mentioned the abundance of deer, births and mentioned the spectacular sunrises and sunsets of the large sky of Texas.

He became a shaman and took the side of the Indians. When he reached Santa Fe New Mexico (1536) he was punished for taking their side and sent to prison in Mexico City. He was sent to South America, where he discovered the Iguazú Falls in Brazil in 1540.

Indians of the Southwest (the Hopi) (Fig. 35)

For two weeks at the San Francisco Peaks of Arizona, members of the Hopi Kachina cult still perform a ritual ceremony to bring rain, just as their ancestors did hundreds of years ago. The officials preside before the Kachina wearing male and female masks, some with painted rainbows to

Fig. 35. Hopi people with Albert Einstein sharing a calumet pipe of peace. These native Americans hold their Kachina dolls while they dance and blow smoke to the heavens to elicit rain. Hopi House Grand Canyon, Arizona. Museum of New Mexico, Palace of the Governors. Neg. No. 38193 (L.U. Gustav Bucky)

represent rain. After finishing their stamping dances imploring the Gods to bring rain, the Kachina must be blessed with special gifts. The priest or chieftain sprinkles her with holy water and corn meal, representing abundance, and the "tobacco chief will blow smoke on their faces so there will be rain. The Kachinas will depart with the priest praying farewell to the holy spirits."[236]

Jean Rivaut (1562) and René de Laudonnièr and Le Moyne (1564) and the People of the Southeast (Timucua and Other Tribes).

In 1562 the French Huguenot leader Gaspar de Coligni sent Jean Rivaut, a French navigator, with a small expedition to explore the lands of the east coast of the present-day United States.[237] The expedition landed on the shore of the St. John River close to the present city of St. Augustine, Florida. These unexplored lands had been named "Florida" by the Spanish explorer Juan Ponce De Leon in 1513. Jean Rivaut explored the coast and found the land so "rich and desirable" that he established Charlesfort (North Carolina, Port Royal Harbor.[238] He returned to France in order to obtain provisions for this settlement and to inform Coligni about his findings, but he did not stay in France because of the ongoing civil war between the Huguenots and Catholics. Rivaut went instead to live in England, where Huguenots were protected, and he met Queen Elizabeth I and the influential Sir Walter Raleigh, the "Admiral of the Seas." Rivaut told them of the discovery of the new lands and said to them that these lands "lacketh nothing."[239] (Fig. 36)

Two years later, in 1564, a new French expedition under René Laudonnièr, also a French Huguenot, arrived in Florida with an artist named Jacques Le Moyne de Morgues[240] who had been commissioned by Queen Catherine de Medici (from the House of Medici) to map the seacoast of Florida and to make an accurate pictorial and narrative description of the natives. Le Moyne De Morgues's paintings, engraved by Debray, give a pictorial representation of the Timucua Indians. In one of these paintings a man is smoking a pipe and a woman is bringing leaves of tobacco

Fig. 36. Creek zoomorphic stone pipe made in one piece. These pipes have been found on the upper Mississippi River area. Museum of the American Indian. Heye Foundation.

to cure a sick person during this health ceremony. (Fig. 37) Here we have testimony that these people considered tobacco a medicinal plant.[241] The Indians of the Southeast smoked tobacco in pipes.[242] These pipes have been found in the burial mounds of Mississippi.

Eventually the Spaniards overran the French settlement and many of the French settlers were killed; however Le Moyne escaped with a few others and went back to England where he stayed and got married.

(Huguenots could not return to France because of religious persecution by the Catholic Church).

Other Tribes of the Southeast
The Temple-Mound Builders: The Hopewell, Natchez and the Cahokia[243]

These cultures lived close to the Mississippi River, where elaborate pipes have been found dating from 800 AD to 1400 AD. They smoked tubular and angular pipes.[244] The Mound Builders extended from the south (Etheral Mound) close to the Gulf of México and up to present-day Ohio, Oklahoma and Wisconsin. Their most advanced culture was the Cahokia (1200 AD).

The Five Civilized Tribes of the Southeast

The Natchez, who lived close to the Mississippi, cultivated corn as the main crop. They prepared many dishes from corn. They were heavy smokers; the growing of tobacco also was very important to them. Other cultures of the southeast were also pipe smokers, including the Creek, Appalachian, Cherokee and Timucua.

The Appalachians yielded steatite, a stone suited to the manufacture of fine pipes. These pipes have been found at many points in the middle of the southern United States as well as other parts of the U.S., especially along the shores of the Mississippi River, which was used for trade and transportation.[245]

Stone pipes were meticulously carved; the drilling of the stem hole was done before the carving of the pipes to prevent cracking the stone. The bore was started with a very slender, flint drill. It is believed that the rest of the bore was done with copper. The "monitor stone pipe" was made of one piece, unlike other pipes that had a separate stem. Zoomorphic decorative pipes with birds and other animal figures were made.[246] (Fig. 53)

The stones were chosen for their color, resistance to heat, and the ease with which they could be worked into shape. The most common stones used were catlinite, steatite, sandstone, limestone, slate, granite, chlorite, dionite, and even fossil coral.

Fig. 37. 'How they treat their sick.' This Timucua ceremony involves blowing tobacco smoke on the face of the sick (left) while a woman brings more tobacco leaves. A sick man is made to smoke a pipe. By Le Moyne. 1564.

The English Explorers of the Northeast Corridor of America

John Hawkins (1562 and 1578–85) and the People of Virginia

When Queen Elizabeth and Sir Walter Raleigh decided to challenge the Spanish Armada, they sent John Hawkins in 1562 to be in charge of the English fleet. The English were no longer restricted by the Catholic popes because England had become separated from the Catholic Church.[247] After plundering the Canary Islands and the west coast of Africa where they captured slaves that were to be sold in Hispaniola (Dominican Republic),[248] Hawkins passed through the Huguenot settlement of Charlesfort (South Carolina) and observed that the Timucua Indians as well as the French settlers were smoking tobacco. Hawkins stated:

> The Floridians when they travell have a kind of herbe dried [dried tobacco], which with a cane [pipe] and an earthen cup in the end, with fire, and the dried herbs put together, do sucke thoro the cane the smoke thereof, which smoke satisfieth their hunger, and therewith they live foure or five days without meat or drinke, and this also the Frenchmen used for this purpose yet do they holde opinion whitall, that it causeth water and fleame to void from their stomach.[249]

Hawkins here was obviously describing tobacco smoking by the Indians, and also by the French Huguenot settlers who had already become addicted to tobacco.

A 1578 expedition by Gilbert, a cousin of Raleigh, and subsequent explorations in 1584 by P. Amadas and Arthur Barlowe explored the coast of North America and gave a gloomy report.[250] The English were unable to establish a permanent settlement in Roanoke, a land which was later called Virginia by Sir Walter Raleigh in honor of Queen Elizabeth, the "virgin queen," so-called because she was not married. For this and other deeds Raleigh was knighted and proclaimed "Admiral of the Ocean Sea."[251]

In the spring of 1585, Sir Walter Raleigh sent another expedition to establish a colony in the land of Virginia. That attempt failed and Drake took the survivors back home to England. (1586) Among the members of the expedition were scientists, including Thomas Hariot who was his tutor of mathematics. In 1587 another settlement was attempted in Virginia under the governorship of John White, sent by Sir Walter Raleigh. Sir Richard Greenville was in command of this expedition. This attempt also failed, and White returned to England when war broke out between Spain and England. The colony had disappeared, "the last colony as reported by Richard Hakluyt." Raleigh himself never touched these lands.[252]

John White was an artist and made the famous sketches of the Pomeiooc and Secoton Indians of Virginia, depicting the plantations of tobacco (*uppowoc*) as well as the Indians smoking and throwing pinches of tobacco in the water during religious ceremonies.[253] (Fig. 38)

Thomas Hariot, whose book 'A brief and true report of the nue (new) found Land of Virginia' was published in 1588, gives us a complete description of the habits of these people as well as the "marketable commodities" of these lands. One of the commodities, tobacco, he describes as follows:

> There is an herb called uppowoc which sows itself. In the West Indies it has several names according to the different places where it grows and is used. The Spaniards generally call it tobacco. Its leaves are dried, made into powder and then smoked by being sucked through clay pipes into the stomach and head. The fumes purge superfluous phlegm and gross humors from the body but if there are any obstructions it breaks them up. By this means the natives keep in excellent health without many of the grievous diseases which often afflict us in England.[254]

Hariot also mentions that tobacco was highly valued by the Indians and was considered sacred by them, as in all other native cultures. He

Fig. 38. Town of Secota. 'Houses of the Towne of Secota have gardens . . . ' First Depiction of Tobacco Farming, L.L. Tobacco patch. by White 1585. NYPL

writes that during sowing of corn and other seeds, "when they made holy fires they threw tobacco into it to give thanks to the gods. Also when they were fishing they threw tobacco into the water to conjure up good luck."[255]

Hariot and his crew stayed in Virginia for about one year, becoming addicted to tobacco-pipe smoking. When he returned to England, he brought back tobacco, the tobacco plant, the potato plant, the habit of smoking, and the tomato plant, whose taste was not well received in Europe. It was from Hariot that Sir Walter Raleigh learned to smoke the pipe.[256] Raleigh became addicted, and as he was a very influential person, he played an important role in introducing tobacco-pipe smoking to the court and, therefore, to the rest of England. He planted nicotiana rustica, brought to him from Virginia, on his Irish estate. This crop was unsuccessful, however, because the English were already accustomed to the milder and scented tobacco from Spain, nicotiana tabacum, which was prohibited in England but was smuggled in by the Dutch and the Spaniards.

Amadas and Barlow brought three Indians back with them to England. The indians brought their pipe and their tobacco to smoke while on their trip and "could not be without it."

The Indians called the English settlers "cutthroats" because they constantly tried to steal the Indians' lands either by war or treachery. Peace was offered, but it never came to the indians because the colonizers wanted their land, mainly to plant tobacco. Tobacco was very popular in Europe and sold in great quantities by the Spaniards and French. Tobacco was already being cultivated in the Island of Hispaniola by the Spaniards. It "made sense" to the English colonists, therefore, to take the land that was already in use by the Indians, rather than to clear the heavily wooded areas by themselves. This led to the displacement of the Powhatan tribes (Fig. C28) farther and farther inland. The confrontation led to wars in 1622, and by the year 1640 this tribal power of the Northeast had been eliminated forever.[257]

Sir Francis Drake (1578)

Sir Francis Drake, while he was at shore in the San Francisco Bay area in 1578 repairing his boat before continuing his trip of circumnavigation of the globe, observed "civilized" Mexicans (Californios) in the West Coast (California) consuming tobacco.

Drake had just completed the circumnavigation of the world and been recently knighted by Queen Elizabeth when he was sent in 1586 by Sir Walter Raleigh to the new colony of Roanoke, which had been settled the previous year, to bring the settlers back to England. Drake left the port of Plymouth with 25 boats and 2,300 mariners. On his arrival in Roanoke, Hariot, the painter John White and all the settlers decided to return with him to England. The settlers were already addicted to tobacco and brought this bad habit back with them to England. On their way back from Roanoke, Drake ransacked the Spanish settlement of Santa Marta (Colombia),[258] Santo Domingo, and St. Augustine (Florida).[259] They occupied Santo Domingo and remained there for about one month. They took everything they could carry, including the bells of the cathedral and one million pounds of tobacco. A ransom of 25,000 ducats in silver and gold was paid by the government and merchants.[260] When they arrived back in England, the English settlers, including Governor Ralph Lane of Virginia, gave a "spectacle of smoking of tobacco from pipes, which caused a great sensation."

England finally established a permanent settlement in Virginia at Jamestown. The Jamestown settlers were making their own tobacco pipes by 1605. Some of these pipes have recently been unearthed (in 2000) by a group of archeologists of the National Geographic Society. Weymouth mentions a welcome he received from the Indians on his trip to the Jamestown colony at the beginning of the 17th century:

> They gave us the best welcome they could, spreading deer skins for us to sit on the ground by the fire, and gave us of their best tobacco pipes, which was most excellent, and so generally commended of us all to be as good as any we ever tooke being the simple leaf without any composition, very strong and of pleasant sweete taste, they gave us some to carry to our captain whom they called our Bashabe, neither did they require anything for it, but we could receive nothing from them without remuneration."

The Iroquois (N.Y.) and the Algonquin (Wisconsin) Indians (Northeast Tribes)

There is an Iroquois saying: "Think not forever of yourselves, or chiefs, nor of your own generation. Think of continuing generations of our families, think of our grandchildren and of those yet unborn whose faces are coming from the ground."[261]

The Iroquois was a confederation founded in 1575 of six tribes that settled mainly around the present-day Finger Lakes in New York State. This federation was accepted and "sealed" by Chief Medusa Atotarho of the Onondaga tribe, (Fig. C29) which joined with England during the America Revolution.[262] General Washington sent an army commanded by General Sullivan to destroy the Iroquois confederacy because they had taken sides with the English. The proud Iroquois confederacy never recovered from that campaign of extermination.

The European settlers called the Iroquois "the Romans of the new world" because of their sophisticated political and social structures.[263] They were also called "the men of the long houses" or "people of the fire that never dies" because of the communal long houses that they inhabited that were obscured by smoke from fires to warm themselves, as well as the smoke from continuous pipe smoking. They, like many other Indians, were inveterate smokers.[264] The Europeans could not stand the stench of these long houses. The Iroquois' symbolic Long House was guarded by the Mohawk tribe, who were great hunters, fishermen, farmers and fighters. The descendents of some of these tribes, sadly, now run gambling casinos.

The women of the Iroquois were in charge of sowing crops such as corn, squash, beans and tobacco.[265] Tobacco was sacred to them. It was considered to be a gift from the "great spirit," and frequent offerings of tobacco were made to him[266]. The Iroquois, like the natives of present-day Virginia, tossed pinches of tobacco into the air to ask the gods for a good harvest. Peace or war treaties were sealed by smoking the communal pipes called calumet. When a trader came to an Iroquois village, he was received with an embrace and the smoking of a pipe. Smoking symbolized the beginning of a friendship.[267]

George Catlin (1830) The Sioux, the Mandan and Other Tribes of the Great Plains

"The whole white race is a monster which
is always hungry and what it eats is land."
Chiksika, Shawnee
A poem by a conquered Indian Nation (the land was taken to plant tobacco).

The Great Plains were called "the American buffalo lands" by the English settlers. This area remained unexplored for almost two centuries after the first English colonies were established. The people of the Great Plains belonged to many tribes, among them the Sioux, the Cheyenne, and the Pawnee. They were semi-nomadic and lived off the buffalo. They did not stay more than ten years in one place. Some of the tribes cultivated their own tobacco, while others bartered for tobacco with other tribes. What they had in common was that all of them smoked tobacco in long-stemmed pipes. George Catlin, a young painter, left St. Louis in 1830 to go west to paint the tribes of the Great Plains. His paintings provide us with the best depiction of the smoking habits of these tribes during their ceremonies.[268]

Catlin wrote in vivid detail about the hardships he endured to arrive at the sacred quarry of southwest Minnesota at the present town of Pipestone. According to the Sioux, the origin of the Great Spirit occurred in these mines:

> The ancient "Great Spirit" at an ancient period here, called the Indian nation together, and, standing on the precipice of the old Pipestone Rock

> broke from its wall a piece and made a huge pipe by turning it in his hand which he smoked over them, to the North, South, East and West and told them that this stone is red—their flesh, that they must use it for their pipes of peace, that it belongs to them all and that the war club and the knife must not be raised on its ground. At the last whiff of his pipe his head went into a great cloud and the whole surface of the rock for several miles was melted and glazed, two great ovens were opened beneath and two women [guardian spirits of the place] entered them into a blaze of fire and they are heard there yet . . . answering to the invocation of the high priest or medicine man who consults them when they are visitors to this sacred place.

This red stone used to make pipes is known today as Catlinite in honor of George Catlin.[269]

The Calumet Pipe (Fig. C30)

This sacred, highly ornate pipe was called the *calumet,* from the French word *chalumeau,* meaning "a long tube or pipe."[270] This word was used by the French missionaries of the early 17th century—Marquette, Heneppin, and others—to describe the stem of the pipe or reed. Later on the whole pipe was called a calumet.[271] These pipes were made of the red stone from the sacred quarries of the Sioux Indians of Minnesota and had a symbolic and religious value because they formed a link between the Indians and the Great Spirit power.[272] According to Catlin, it was called the pipe of peace but was also the pipe of war. It was the instrument to seal pacts, a passport to obtain free passage through friendly and enemy lands, and a means of communicating with the great spirits, or *Wakan Tanka,* of the Sioux tribes. Marquette in (1673) tells us that for these people there was nothing of more importance that the calumet pipe. They gave such importance to this pipe, most likely, because it was the censer where tobacco was burned. Tobacco had a sacred, divine significance for the Sioux, as it did for all other tribes of America. Not only was it a way to communicate with the Great Spirit, but that spirit was present in the pipe itself.

Each tribe had its own type of calumet pipe. They were beautifully carved and gave the tribe its identification. The stem was decorated with colorful feathers and with the hair of the horses of defeated enemies. Red feathers only adorned the calumet pipes of war. These pipes also were used by the shaman during healing ceremonies. Some calumet pipes were used for enjoyment, but they were not sacred, ceremonial pipes. The sacred stone quarry of southwest Minnesota where the rock for the pipes was obtained was neutral ground, respected by all tribes. No fighting could occur there.

The Sacred Calumet Pipe and the Great Spirit

"Wakan Tanka [the Great
Spirit] when I pray to him Hears
Me, whatever is good, he grants
Me."
Teton Sioux Dance Song

According to a Sioux legend, when the Great Spirit created the spirits of nature he conferred certain powers to each animal, different from the powers given to man. He created a special plant—the tobacco plant—for those who were weak. When tobacco smoke was smelled by all the spirits, they developed a strong craving for it and they wanted to give up their powers for the power of the tobacco plant. The Great Spirit denied this plant to them because it was given only to man, but he left it to man to decide to share or deny this power to the spirits at their discretion.

Vera Louise Drysdale gives an account of the seven rites of the Ogolala Sioux based on Black Elk's account, "The gift of the sacred pipe." This depiction is an interpretation of what Chief Black Elk might have seen during the ritual ceremony in the Sioux cosmology:

> The calumet pipe for them was also a symbol of the sacred rites of the Sioux, the dance every year around the sacred tree of the world that is planted by them every summer to celebrate the sun dance at which time they celebrate the life principal. The white buffalo woman gives the pipe to the Sioux so that they might "send their voices" to Wakan Tanka the great spirit, the seven rites of these people are recited.[273]

"An Indian believed that smoke was to pray.
A portable altar, the pipe was a combination of important symbols.
The pipe stone bowl represented the Mother Earth. The wooden
Stem represented all growing things. Red porcupine quillwork
Was a symbol of magic power. By mingling his breath with sacred
Tobacco and divine fire, the smoker (i.e. the worshipper) put himself
In tune with the universe."
D.H. Miller, *Custer's Fall: The Indian Side of the Story*[274]

Man, therefore, was thankful to the Great Spirit and made offerings of tobacco to him, throwing it in the rivers or the air, but mainly burning it in their ceremonial calumet pipes so that the smoke would go to heaven and they could communicate with him.

The Crow Indians

The Crow were semi-nomadic hunters living off the buffalo, corn, and a few other plants.

They had little tobacco to spare and used it mainly for religious rituals.

All indigenous people of North, Meso, and South America had differences between their cultures, but all of them share one feature in common—they were addicted to tobacco. They had an undeclared "brotherhood of the leaf," using tobacco in religious and civil ceremonies, for medicinal purposes, and for personal enjoyment. They used tobacco in different forms—in pipes, cigars, snuffs, and chewing.

The Sioux chiefs were the custodians of the pipes, which had different "powers." The pipes were used with pride to display power, but with discretion. They mixed tobacco (nicotiana rustica) with *shongsasha* (red willow bark).[275]

In the mid-19th century, during the California Gold Rush, more than 200,000 colonizers invaded the Great Plains on their way to California. The Indian lands were taken by force, by treachery, or by peace pacts in which the calumet was used, but never respected by the white man, who used the symbolic calumet pipe to sign treaties only when it was convenient to him. By that time the value of the calumet as a symbol of peace, or its importance in sealing agreements or pacts, had lost its value. At the end of the 19th century, the Sioux were displaced from the sacred quarries from which they had made the calumet pipes. The white settlers had succeeded in taking their lands.

The Pawnee

The Pawnee carried with them at all times the sacred bundle hanging from their necks that contained a pipe, pigments, tobacco, and corn. The bundle was passed from father to son.

"The Sacred Bundle"
We value our medicine bags so highly
that we would not part with them while
life endures.
Some of us did, and
This proves to be the source of many

Heavy calamities.
Winnebea, Sauk warrior

The Great Lakes and the North Mississippi River

Samuel Champlain (1603–1620)

Samuel Champlain was the founder of Quebec and is called the "Father of the New France (Canada)." He first entered New France in 1603 as an artist, mapmaker, and explorer. Champlain first explored the north coast of the present-day U.S., but left those lands because of the hostility of the people in the area of present-day Manhattan. He made the first maps of what is now New England, and developed good relationships with the Indians. During his exploration of the St. Lawrence River, he and his crew lived with the Huron people. They integrated into their ways of living and married among them, and from these intermarriages a new ethnic group was born (the Metis). Champlain took the side of the Huron against their enemies, the Mohawk (Iroquois).

Champlain had first noticed smoking by the settlers of Borinquen, present-day Puerto Rico, while navigating on a Spanish ship.[276] While exploring the area of present-day Quebec along the St. Lawrence River, he observed that one tribe of Huron, the Wyandot, was entirely involved in the cultivation of tobacco. For this reason they were called "the tobacco nation," and their people were named "the tobacco people." They also traded in pelts, corn and wild rice.[277]

Another observer, Jean de Brebeuf, wrote of the Menonidee Indians in 1636: "They believe that there is nothing suitable as tobacco to appease their passions, that is why they never attended a council without a pipe or calumet in their mouths. The smoke, they say gives them intelligence and enables them to see clearly throughout the most intricate matter." They also used it in their "sweat baths" during epidemics of smallpox brought by the Europeans. The vapor in the sweat bath was created by throwing water with herbs and tobacco onto red hot stones. This habit was (and is) common also in Mesoamerica (México).[278]

Tobacco was believed by these tribes to be a gift from the Great Spirit (as the Europeans called it). To thank the Great Spirit they made frequent offerings of tobacco. A large calumet of red pipe stone, with a long, ornamented stem with bright-colored strings was filled with a mixture of tobacco and *kinneckinick* ("that which is mixed," tobacco mixed with the inner bark of dogwood, with sumac leaves and oils as a binder). When a wild rice harvest was about to begin, a few pinches of dried leaves of tobacco were tossed in the waters to ensure ample picking. When a man encountered a large obstacle like a rock or tree trunk in the forest, "he would sprinkle tobacco on the ground to ward off wrath by the spirits that might bring about misfortune." These people believed that spirits were everywhere, watching them.

The Natchez got "intoxicated" with tobacco, mixing the tobacco with sumac to sweeten it (as did other tribes of North America), sucking the smoke in deeply ten or twelve times, and keeping the smoke in for a long time until they got into a state of trance.[279]

The Missionaries—Jacques Marquette and Louis Joliet (1666–1673)

"To a Jesuit Missionary by a Native Algonquin"
This is well, "Black Gown," that
thou cometh to visit us. Take pity
on us:
Thou are a Manitou, we give the
tobacco to smoke . . . Let the earth
give us corn and the river yield
us fish; let not disease kill us any
more or famine treat us any longer
so harshly. "
Algonquin elder (a Huron) upon seeing the first Jesuit in 1670
(Alvin M. Josephy Jr.)

Jacques Marquette was a Jesuit missionary who arrived in Quebec in 1666 and studied the Indian customs and languages. He founded a mission in the present area of Michigan. Marquette was commissioned by Louis Frontenac, Governor of New France, to explore the Mississippi River. Later they descended to present-day Arkansas. He went down the river by boat and was the first white man to live in the area of the present-day city of Chicago, where he lived and observed the customs of the Illinois Indians, including the custom of smoking the calumet pipe. This pipe was sacred to them during peace or war and was respected by all as a passport through enemy lands. Marquette wrote in the early 17th century: "It was respected more than the crowns of kings in Europe. It was mysterious, it was god of peace or war, the arbiter of life or death to be shown when passing through the enemy land and when displayed in battle was sufficient to stop a fight. It was considered mystical because of its association with the sacred tobacco.[280]

Louis Hennepin and René Robert Cavelier (1679–1682)

Louis Hennepin was a Franciscan missionary who went with the celebrated explorer René Robert Cavelier de La Salle to the region of the Great Lakes in 1679. They explored the upper Mississippi River and were captured in April 1680 by the Sioux Indians. By living with them and

Fig. 39. Father Hennepin's depiction of the Calumet Pipe. 17th c. He explored the upper Mississippi region. NYPL

sharing their customs for some time, Hennepin (Fig. 39) and Cavelier learned to appreciate the significance of the calumet pipes in their religious and civil ceremonies. When Catlin explored these lands and came into contact with the Indians in the 19th century, he found that they were well aware of the European settlers. Hennepin reached and lived in the area of present-day Minneapolis where he made a painting of a pipe.[281]

The Lewis and Clark Expedition, The Missouri Expedition 1804–1806

President Thomas Jefferson negotiated the purchase of the Louisiana Territories from Napoleon Bonaparte in 1803. Bonaparte at that time had his hands full with wars in Europe.[282] In order to determine the boundaries of these territories, Jefferson sent Meriwether Lewis and William Clark on an expedition. (Fig. C31)

Meriwether Lewis, an intellectual and experienced woodsman, was a Virginia neighbor of President Jefferson and became Jefferson's private secretary in 1801. The President trusted Lewis and placed him in charge of the Missouri River expedition. Lewis later became the first governor of Louisiana.[283]

William Clark was a Virginia-born Kentuckian and brother of George R. Clark, a well-known soldier of George Washington's army. William Clark was a frontiersman, a good woodsman, and an experienced soldier who skillfully observed the importance of tobacco to the Indians and used it as his "passport" in his travels. The people called Clark "the red-headed chief." After the Missouri River expedition he was appointed as superintendent of Indian Affairs. Clark wrote in detail about the expedition and documented it with sketches and maps.[284]

Among the articles purchased for the expedition were 12 tomahawk pipes and 130 rolls of tobacco (63 lbs) to this another 136 pounds of tobacco were bought at St. Charles in May 1804. Only seven members out of 29 members of the expedition did not smoke (Woodger Elin, pp. 342–343). Curiously enough, among the medicine that they took along were 15 lbs. of pulverized cortex of Perú, julap, rhubarb, ipecaccan, camphor, opii Turkish laudanum, along with a small packet of surgical instruments, small lancets, and 50 dozen bilious pills of Benjamin Rush. Most medications were from native plants of America.

Lewis received full instructions in many matters before the expedition. He and Clark used tobacco as a way of passage through territories occupied by friendly and hostile tribes, "for these tribes use it as a way to make a deal."[285] (Fig. 40) The Indians had two elements of diplomatic etiquette: one was smoking tobacco, and the other was gift giving. Lewis and Clark smoked the pipe with many tribes as a symbol of peace and friendship while on their journey. They spent two and a half years exploring these lands and they recorded the smoking habits and other customs of the inhabitants.[286]

In June 1804, after they had spent the winter with the Mandans and while they were marching towards the Continental Divide of the Rockies

Fig. 40. A Cheyenne smoking a calumet. Lewis and Clark brought along pipes as a means of passage. Their sweet tobacco was greatly appreciated since native tobacco like the Arikara's was too strong. Recorded on Indian cloth. Werner Forman Archives. Private collection.

west of Montana looking to trade for horses for their expedition, the expedition encountered sixty mounted Shoshonee warriors. Fortunately, Lewis and Clark were in the company of the Frenchman, Charbonneau, and his wife, *Sah-cah-ga-we-ah* (or Sacajewea). The warriors talked to Sacajewea and then met the explorers in peace. The chief of the warriors turned out to be the brother of Sacajewea. She had been kidnapped by the Hidatsa while still a child. Sacajewea was very important to the expedition because, according to the Indian tradition, "a woman with a party of men is a token of peace."

The warriors embraced Lewis and Clark very affectionately, applying their cheek to them and saying ah-hi-e, ah-hi-e ("I am much pleased, I am much rejoiced"). Lewis and Clark lit the pipe and gave them smoke. In his diary Clark described what followed: "After sitting in a circle and pulling off their mockersons [moccasins] they started smoking and passing the pipe, meaning friendship and a sincere commitment to their word, meaning that by removing their shoes they would go forever barefooted if they were not sincere."[287]

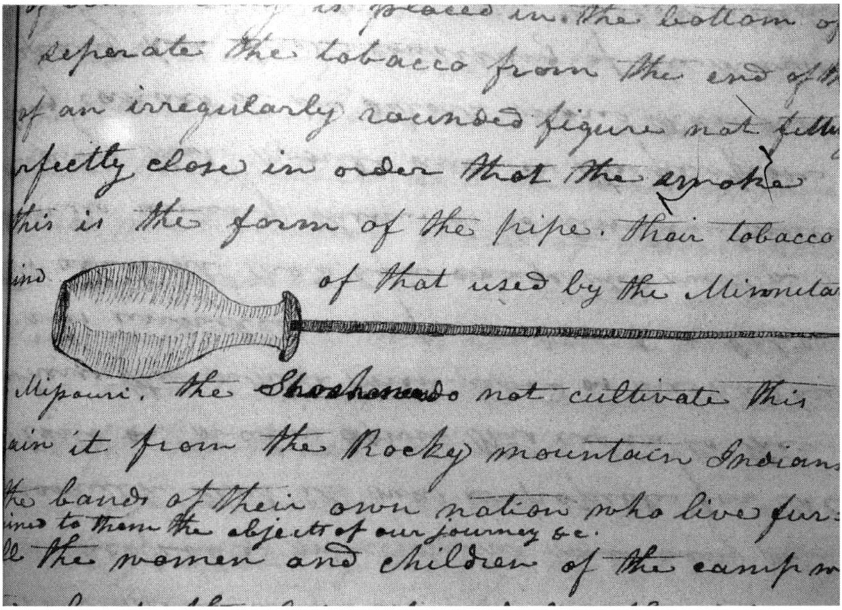

is placed in the bottom of
separate the tobacco from the end of th
of an irregularly rounded figure not fitt
pfectly close in order that the smoke
this is the form of the pipe. their tobacco
ind of that used by the Minnetar
Missouri. the Shoshones do not cultivate this
ain it from the Rocky mountain Indians
the bands of their own nation who live fur
ined to them the objects of our journey &c.
ll the women and children of the camp w

Fig. 41 Arikara pipe. Clark drew this pipe while going through their territory. From Clark's diary book. Smithsonian Museum.

Lewis and Clark also had the help and company of Twisted Hair, a Nez Perce chief in his mid sixties. Twisted Hair's help was extremely important on the last leg of the expedition over the Bitterroot Mountains (the Continental Divide). On the trip down the Columbia River other chiefs came with him, including Tetohorsky, who reassured the natives that the white men were on a peace mission.

When Captain Lewis encountered the Lakota on September 28, 1804, they demanded tobacco and a flag. Lewis refused at first, but then he gave in for his own convenience and, as he wrote, "I threw a carrot of tobacco to the chief." Giving tobacco away became a commonplace throughout the journey. All of the people the expedition encountered valued tobacco, especially the Virginia tobacco Lewis and Clark had brought with them. (The tobacco grown by the Lakota was nicotiana attenuata and nicotiana quadrivalis which was much stronger than the Virginia tobacco, nicotiana tabacum, imported from the Orinoco.) The explorers had had the foresight to pack plenty of tobacco as a means of obtaining passage through the natives' territory and to barter for goods such as corn, pelts and even horses.

The Nez Perce eagerly accepted tobacco and other commodities in barter for horses at the Columbia River. The Nez Perces used tobacco during their burial ceremonies.[288] The Crow and Blackfoot cultivated only tobacco; no other crops were sowed.[289] One of the tribes, the Arikara, cultivated tobacco (N. quadrivalis) not only for their own consumption, but also for barter. Clark sent President Thomas Jefferson a gift of Arikara tobacco from Fort Mandan, North Dakota. Jefferson then cultivated it at Monticello. In his diary Clark drew the type of pipe that the Arikara

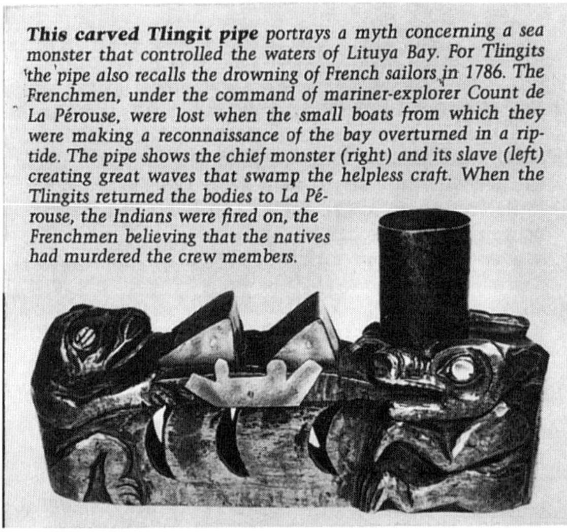

Fig. 42 Photography taken from *America's Fascinating Indian Heritage,* a Readers Digest book.

smoked.[290] (Fig. 41) At the burial of a Comanche Indian, a handful of tobacco was buried with him in order to ensure him of a peaceful death.

On April 3, 1805, many articles obtained during the first part of the expedition were sent down the Missouri River to St. Louis, and from there to President Jefferson in Washington. These included skeletons of never before seen animals, plants, minerals, and a "carrot of Ricara Tobacco;" these people were enemies of the Mandan.

An April 20, 1805, Clark's diary entry described the "remains of some indian hunting camps near which stood a small scaffold of about 7 feet high on which were depicted two dog slays [sleighs] with their harnis [harness]." Underneath this scaffold lay a human body rolled in several dressed buffalo skins. Near the body was a bag containing sundry articles belonging to the deceased, including a pair of moccasins, some red and blue earth, beaver's nails, instruments for dressing the buffalo skin, some dried roots, several plants of the sweet grass, and a small quantity of Mandan tobacco. "Near the scaffold I saw a large carcass of a dog, not yet decayed, which I supposed had been killed at the time the human body was left on the scaffold."

Lewis and Clark often smoked the pipe of peace while crossing the Indian lands as a sign of peace and friendship. Jefferson Medals of peace were given to the chiefs during these encounters, and the tomahawk pipes were highly valued by them.

While at Fort Clatsop, Oregon, on December 23, 1805, at the end of their journey, they divided up what was left of their tobacco, which amounted to twelve carrots, "one half of which we gave to the men of the party who used tobacco, and to those who doe not use it we make a present a handkerchief."[291] After exploring the sandy beaches of the Pacific, Clark found the two chiefs smoking on his pipe.

The party stayed at Fort Clatsop from December 8, 1805, to March 23, 1806.[292] The fort was divided into two large rooms, one for the enlisted men and the other for Lewis and Clark, Charbonneau, his wife Sacajewea, and their young son Jean Baptiste. A smaller room was the smoke house.

The Indians of this area were the Clatsop and the Chinook. While the party was making repairs to the boats, "they had a passing smoke with Twisted Hair and Tetohorsky." Lewis described how the Chinook smoked:

> They are excessively fond of smoking tobacco. In the act of smoking they approve swallowing it as they draw it from the pipe, and for many draughts together you will not perceive the smoke which they take from the pipe in the same manner they inhaled it in their lungs until they become surcharged with this vapor when they puff it out to a great distance through their nostrils and mouth. I have no doubt the smoke of the tobacco in this manner becomes much more intoxicating and that they do possess themselves of all it's virtues in their fullest extent; they frequently give us sounding proofs of its creating a dismorallity of order in the abdomen nor are those light matters thought in delicate in either sex, but all take the liberty of obeying the dictates of nature without reserve.

The Chinook overpriced their fish and roots, and some of "these fellows were found assuming and disagreeing but we smoked with them with every attention and friendship." Clark got very upset, however, when he found that "one of these fellows had stolen my Tomahawk pipe which they were smoking with." On March 23, 1806, they began their journey back towards the Missouri River.

On March 7, 1806, when their supply of tobacco ran out, Gass noted that twenty-six of the men were now without anything but the bark of crab trees to satisfy their chewing and smoking habits.[293]

California and the Great Basin

The first indigenous people of southern California to see the white explorers passing on their way to Asia were those living close to the coast. In 1578, Sir Francis Drake gave an account of these people while circumnavigating the world. He noted that they were smoking tobacco, and said, "the people came everyday with feathers and small bags filled with tobacco."[294] These Indians smoked tobacco in tubular pipes. They planted nicotiana rustica, but also knew nicotiana attenuata and other nicotiana varieties, some of which grew wild.[295] After the first mission was established in California in 1769, the lives of these Indian tribes completely changed. They became enslaved to the white masters.

The Northwest, Canadian Coast, and Alaska.

Among the Klamath people both sexes smoked tobacco, mainly in tubular pipes. It was not used for ceremonial purposes. They also chewed tobacco powder mixed with lime. The Karuk only smoked tobacco (n.attenuata).[296] When General Vallejo settled in the present-day Sonoma Valley in 1839, he brought artisans with him, including brick makers, shoemakers, tanners, a cigar maker and physicians. At that time San Francisco had three doctors and one cigar maker.[297]

Vitus Jonassan Bering (1728)

The fever for exploring the new continent had gripped all of Europe. Vitus Jonassan Bering was a naval officer and explorer in the service of Tsar Peter the Great who was sent on an expedition from Russia to explore the northwestern part of the continent.[298] While exploring the coast of Kayack Island (Alaska) in 1728, Bering sent a German naturalist, George Wilhelm Steller, to explore a nearby Indian settlement. Steller found a hut that the Indians had abandoned in a hurry, leaving behind clothes, an iron knife, and tobacco.

These Indians did not grow tobacco; they obtained it in trade with people of eastern Siberia.[299] They smoked tobacco in small pipes. Only a small pinch of tobacco was used since it was very scarce and could not be obtained easily.[300] It was a very strong type of tobacco (possibly nicotiana rustica). They placed dog hair at the bottom of the pipe as a filter, and passed the pipe around. The individual with the highest position in the tribe smoked first and then passed it to others, who refilled the small pipe (which was only good for a few puffs). Interestingly, the Japanese and Chinese at about the same period smoked tobacco using similar pipes with a small bowl. This suggests that not only tobacco, but pipes also were traded from Asia to Alaska. There are similarities between the Eskimo pipes and the Japanese and Chinese pipes. The Japanese were dainty in smoking their pipes; the *kisseros* were tiny and held only a small pinch of the finest cut and mildest tobacco. From each pipeful only one deep whiff was inhaled. The burning remnants were taken out and a fresh charge was inserted for a second smoke.

According to Captain Corcoran of the Katzebue expedition to Alaska in 1816, some of these Indians were so fond of tobacco that they "smoked it, chewed it, and sniffed it at the same time." Corcoran said that "smoking is their favorite habit. They indulge as long as their tobacco would last. Parties got together to enjoy the smoke in all kinds of pipes; tobacco was not used in religious rituals. The greatest enjoyment that they had was to inhale the whole content of these small pipes in one breath and those who failed to do so were made fun of."[301]

Count Jean Francois Galaup de la Pérousse and the Tlingit of Alaska (1785–1788).

In 1778 the French explorer Count François de la Pérousse was commissioned by Luis XVI while at Versailles, France, to explore Alaska and noted that the Tlingit of Lituya Bay made elaborate carved pipes of stone or whale bone. Fish was the main staple of their diet and they had no other crops except tobacco.[302] The Tlingit were active traders and experienced navigators, using canoes that often were beautifully carved. To commemorate the shipwreck of la Pérousse on the shores of Lituya Bay, the Tlingit carved a pipe depicting the incident.[303] On the right side of the pipe is the monster that caused the shipwreck, and on the left side is the slave that is at the monster's mercy. In between are large waves that swamp the helpless craft. (Fig. 42)

According to Derek Wilson "his ships Baussole and Astrolobe had both come to grief on the vicious reefs of Vanikovo and atoll in the New Hebrides."

Parkman's 1846 Journey on the Oregon Trail

Francis Parkman, who led the 1846 and 1849 Oregon Trail expedition and is known for his seven-volume history of the Anglo-American struggle for North America, noted the importance of pipe smoking to the native Pawnee. When the Pawnee were given half a pound of tobacco by Parkman as a sign of good will, they accepted this gift with great gratification.[304] The Pawnee used to mix tobacco with shongsasha (red willow bark), putting this mixture in their red bowls. They shared the pipe, passing it around, each one enjoying a few puffs.[305]

Alexander Ross

Alexander Ross spent the years 1810 to 1825 exploring the Northwest while trading for furs. He mentions how important tobacco was for his trade: "They smoke as an introduction to all negotiations and only when they were finished they began business." While Ross was making his way toward the Rockies, he encountered the "Snake Nation" (Nez Perce), who cultivated their own tobacco. This tobacco came from a low plant. "They rubbed it among their hands and pound it until it was having a taste resembling green tea but they appreciated the white man's tobacco as well."[306]

At the end of his expedition, Ross met the Chinook Indians of Oregon and Washington. Ross mentioned that he was always well treated by these people and he had brought along plenty of tobacco to share with them. Ross used tobacco as his passport, just as Lewis and Clark did. One time Ross encountered a group of 500 well-armed and menacing warriors who were ready to attack. He shared the ceremonial pipe with them, and when the ceremony was over they were amiable. Each one got a head of tobacco, and each one of the "motly" group was given a single leaf. They had never been treated like this before, and with one voice they all shouted, "*Haugh owe yea ah, Haugh awe yea ahi!*" ("Our friends, our friends!")

Another explorer of the North Pacific, Gabriel Franchiese, mentioned that the Chinook Indians were not interested in alcoholic beverages, but they were very fond of smoking tobacco until they were intoxicated.[307]

PART II

The five Coats of Arms of the European colonizers of the new continent of America. American Heritage Frontispiece, 1963.

TOBACCO TO SPANISH GOLD
The Spaniards converted tobacco to gold.
Tobacco brought riches to Europe,
and slavery, misery and death to America.
Tobacco proved more addictive than gold.

Nebur Azeporo (1930–)

8.

INTRODUCTION OF TOBACCO INTO EUROPE AND SLAVERY INTO AMERICA

In 1493, Columbus had to leave behind 40 sailors in Hispaniola; because he had only two caravels to go back to Spain. The Santa María had sunk in Hispaniola.

The Spaniards had abused the Taíno men, forcing them to work and till the land. They were not used to husbandry Their women were also abused and raped, and for this they paid dearly.

All the Spaniards were murdered. This would be the beginning of poor relations between the Europeans and the Indigenous people of America for centuries to come.

The Spaniards that came in subsequent trips to the New World, came entirely for selfish reasons, not to christianize the people or improve their primitive conditions, but to exploit them and, enslave them.

When the Taíno had been decimated due to overwork, smallpox, measles or other diseases, the Spaniards brought slaves from Africa. This began as early as 1501 when the first black slave was brought to America.

Charles V, King of the Holy Roman Empire, officially endorsed slavery in 1518.

There were few voices that expressed the conscience of Christianity and the Spanish people. One of them was Antonio Montesino who had been sent to christianize the Taíno. He protested and reacted with this vitriolic speech (excerpts) in 1511: "I am the lonely voice of Christ preaching in the desert, in this Island (Hispaniola).

"You are in mortal sin,
You have abused and been unjust to these people
You are killing them.
Don't they also have a soul?
Are not they children of God?
Can you love them like you love yourselves?
Don't you have feelings?
Don't you understand?
Are you out of your senses? Are you mad?
Yes, you are! All this because of your greed for gold.
You are heading for damnation."

These words fell on the deaf ears of these settlers and of those who later settled in America during the next four hundred years.

Slavery, misery and death was brought to the indigenous people of America and to the blacks of the African Continent, only because of the greed of sugar, silver, tobacco and gold.

9.

EUROPE TAKEN BY STORM

"The Old World was sure forlorn wanting thee!"
Charles Lamb (1775–1834) on tobacco

The magical weed appeared in the old continent at the right time—during the Renaissance. It was a time of great excitement. Man was liberating himself from the chains of religious dogma and tradition of the dark ages. Europeans wanted a change, new ideas, new experiences. They were ready to experiment, and found tobacco unique, inspiring, soothing, and sublime!

Fifty years after Columbus—a Renaissance man—first encountered the New World, tobacco appeared at the Portuguese court. It had been previously introduced to Europe by mariners—the 'nicotine missionaries' returning from the New World. It was after its appearance in Lisbon that the leaf spread throughout Europe, permeating all levels of society with a rapidity that has yet to be equaled by any other foreign commodity. Another striking and unusual fact about tobacco during this introductory period in Europe was that it was smoked by all classes of society.

One way Europeans obtained tobacco was by assimilation. The great Spanish physician Nicolás Monardes[1] studied tobacco by cultivating it himself and then extolling its virtues in his history of the New World's medicinal plants, a sort of medical bible for European physicians for at least two centuries. Monardes listed at least twenty medical disorders or complaints that tobacco could cure. Other popularizers of tobacco included Sir Walter Raleigh and Jean Nicot, the French Ambassador to

Portugal, who sent seeds he obtained in Portugal to the French court. It is from Nicot's name that the word "nicotine" is derived.

Since pre-Columbian times tobacco had been used by the people of America in many ways. The Spaniards in the 16th century smoked it mainly in the form of cigars, because that was the most common way it was smoked in the West Indies. The Portuguese smoked it in pipes, but also in the form of cigars as was done by the people of Brazil. The English started smoking it in pipes; that was the way the people of the North American smoked it. Pipe smoking was accepted not only in Europe, but also in Asia during the 17th century. It became chic for the aristocracy to smoke tobacco in pipes and long cigarette holders. At the end of the 17th c the world was in smoke.

In the 18th century came the new fad of inhaling tobacco powder, *rappee* or snuff. Among its practitioners was Napoleon Bonaparte, who was a heavy user and collector of snuffboxes. In the 19th century, people switched to cigar smoking, and in some countries, such as the United States, chewing tobacco and spitting in public was a part of the scene by the end of the century. The development of the cigarette and the match box at the end of the 19th century changed everything. Tobacco use became more convenient. Cigarettes could be put in a pocket or a purse, and could be lit at any desired moment for a short time and without interruption of any activities. Life at the end of the 19th century during the industrial revolution was becoming hectic. The convenience of a fast cigarette puff was the answer. Entrepreneurs developed—first in Europe, and then in the United States—sophisticated machinery to make a large number of cigarettes and matches that could be easily and safely lighted. They also developed and mastered the art of advertising this product for general consumption, which eventually developed into one of the most successful advertising enterprises. The tobacco industry carries a lot of clout in our modern-day societies, and influences government policies in many ways.

10.

INTRODUCTION OF TOBACCO INTO SPAIN, PORTUGAL AND FRANCE

"O noble tobacco. To be praised
above all weeds which in East and
North/ in West and South the
Celebrated sun causes to shoot out
of the earth. How much praise
your strength and your virtue is
worth"
Jan Jansz Starter (1594)

Spain, Portugal, and France

Rodrigo de Jerez, one of the two people that Columbus had sent on a diplomatic mission to find the Great Khan, had already become addicted to tobacco while he was exposed to it in the Islands of Cuba and Hispaniola in 1492–1493. It is thought that Jerez, after his return with Columbus to Spain, was the first European to smoke tobacco in Europe in the form of a cigar (in 1493). "The people of his native town of Ayamonte [a Spanish town close to Palos de la Frontera], when they saw smoke coming from his nostrils and his mouth, were terrified, thinking that the Devil had gotten possession of him. He was referred to the Inquisitor, remaining in prison for seven years until the habit of smoking was very common."[2]

Oviedo brought tobacco to Spain in 1519, and it has been said that Columbus himself and Cortés brought it back also, but it is not proven. (The same has been said of Pizarro's soldiers, that they brought coca leaves to Spain).

Francisco Hernández de Toledo (1514–87), a physician and botanist sent by Phillip II to investigate the new plants of México from 1571–1577, found about 1,200 medicinal plants. He sent his reports to Spain and described the beauty of the tobacco plant.[3] The people of Spain started cultivating tobacco in their gardens mainly because of the beauty of the plant.[4] Dr. Nicolás Monardes, a physician in Seville, Spain, was given the task of investigating medicinal plants brought from the New World, among them tobacco. In 1565 it was introduced as henbane.

At the beginning of the 16th century, there was little demand for tobacco in Europe. It was only grown in small parcels for personal consumption in Hispaniola (Dominican Republic) and Brazil, but it was imported in small amounts to the ports of Europe as a commodity. The Portuguese started selling tobacco in Lisbon in 1535. One tobacconist was Antonio Baço, *Mercador* (merchant). The Portuguese called tobacco "*Fumo*." The authenticity of this has been challenged,[5] but it is still called by that name in some parts of Brazil today.

The habit of smoking took hold in the ports of Europe after a short time. The new smokers made a special request to the sailors traveling to the New World; they wanted neither gold, nor silver, but a bushel basket of the "Indian Weed." The demand in Europe increased exponentially, and many settlements in Hispaniola were planted with tobacco by the Spaniards. On the coast and near the Orinoco River, tobacco was planted by the Portuguese and the Spaniards, and it was cultivated in large parcels in Sao Vicente in Brazil (1534). At the end of the 16th century, the Spaniards and Portuguese were selling the leaves of tobacco in all ports of the world. The demand was endless.

The Spaniards brought tobacco to the Philippines in 1575 via Acapulco, México. Before then, Ferdinand Magellan (Fig. C35) supposedly had introduced tobacco to those islands[6] during the first circumnavigation of the world (1519–1522). Magellan had sailed past Brazil and Argentina to the tip of South America (Patagonia) and Tierra del Fuego, to the Pacific Ocean. The first landing was at Guam before getting to the Island of Cebu in the Philippines. (Fig. 43) Later, tobacco was brought from the Philippines to China and Tibet. Magellan was killed in Mactan, Mindanao, Philippine in 1521.

In 1772, Mindanao, South of the Philippines exported large amounts of tobacco. (Captain Cook's second exploration. Book by Dampier.)

In the 15th century, Portugal had a large trade with India. This trade increased after the discovery of the New World. The Portuguese and other European explorers spread the use of tobacco to India, the Far

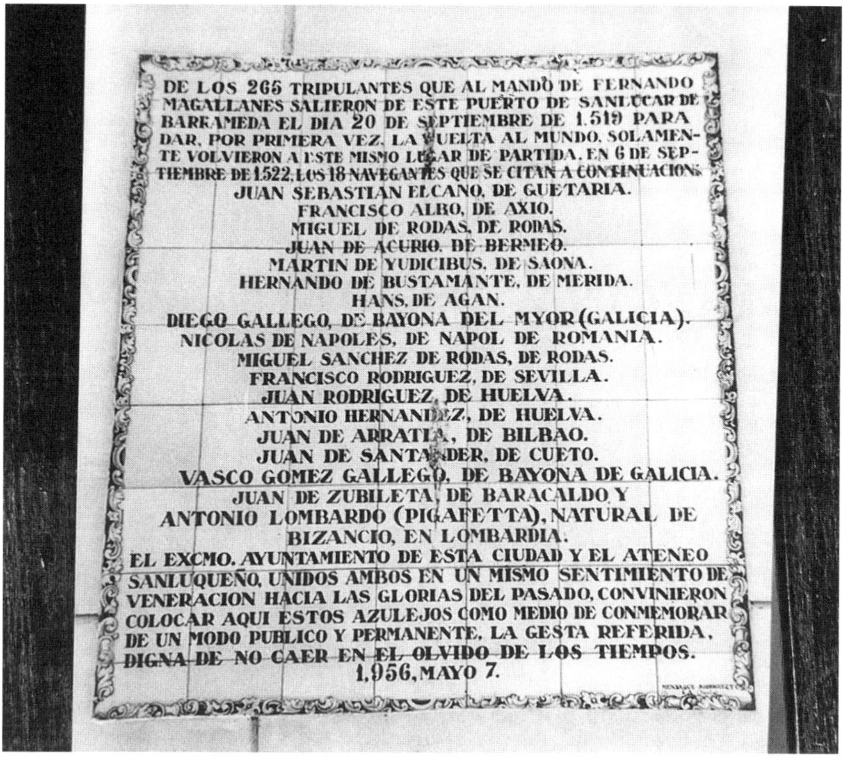

Fig. 43. Survivors of Magellan's Molucca expedition. Magellan, commanding the Trinidad, left with five caravels. Only one ship made it back to San Lucar de Barrameda three years later under the command of Juan Sebastian Elcano, of Magellan's 265 sailors, only eighteen survived. Plaque on the wall of the Old Ayuntamiento (Town Hall) in San Lucar de Barrameda, Spain. Photography by Roger Hagan.

East, and the coasts of Africa which is why the Africans called tobacco "*Bortugal.*" The Portuguese had learned from the people of South America how to smoke tobacco in pipes. They took not only the dry leaves of tobacco, but also its seeds, as far as Macao.

Portugal had become a major power by the middle of the 16th century and had colonies in Asia (Ceylon), (Thailand), Africa, Brazil and India (Goa). Lisbon and Seville became the main centers of trade in Europe. It was only natural, therefore, that Portuguese sailors would spread the habit of smoking throughout the world as far away as Japan. Merchants from all European cities came to Lisbon and Seville to buy the latest imports from all over the world, including Damascus silk, spices from India, and silver and gold from the New World. Thus, the use of tobacco became known to many merchants as well as physicians.

11.

MEDICINAL USE OF TOBACCO

France and Spain

Some of the first observations of the "medicinal" use of tobacco were made in 1523 by Giovanni da Verrazano. Verrazano mentioned the medicinal use of tobacco leaves by the Indians of South America, he passed this information on to King Francoi I of France.[7] In 1543 there is mention of tobacco already being recommended for medicinal uses in Salamanca, Spain. The belief in the medicinal properties of tobacco was shared by herbalists, monks, lay people, and physicians such as the Spanish doctor Nicolás Monardes. Europe was "hypnotized" by this herb. Pharmacists were using it as a dried leaf and preparing it in all kinds of combinations, depending on the particular illness that tobacco was supposed to cure. The high dignitaries of the Catholic Church sent the tobacco plant with their recommendations and "blessings" to wherever they had a monastery and recommended its cultivation.

Jean Liébault, a botanist, and his brother published their book *La Maison Rustique* in 1571.[8] Together with Nicolás Monardes (also known as Nicolai Monardis or Nicolò Monardes) (Fig. 44) a physician of Seville from the University of Salamanca, they were instrumental in listing all the ailments for which tobacco was supposed to be of benefit and established a pharmaceutical code of how to use it. Monardes (1515–1588) acquired great fame in Europe for his "contribution" to the armamentarium of medicine with his book *Dos Libros* (Two Books) in 1569–1571. "The herb, commonly called tobacco," Monardes wrote, "is a very efficient herb, known among the Indians, chiefly those of New Spain . . . healing themselves with it to the great benefit of all . . . " According to the Indians, it was good for unguents (ointments), powders, gargles, emetics, enemas,

Fig. 44. Nicolás Monardes. First and main proponent of tobacco as a medicinal plant, in Europe. (Corti)

cathartics and other concoctions used for nearly every disease. If a patient had a cough, he had to smoke tobacco, inhaling deep; for a running sore, a local application of a tobacco leaf; for a toothache, a mouthwash of a tobacco solution; for pains during labor, a hot leaf of tobacco on the navel; for delirium, tobacco was blown in the nostrils; for headaches, a sniff of snuff tobacco powder, and when some body had drowned, resuscitation was attempted giving an enema with tobacco." Smoking supposedly was good for arthritis and for every real or imaginary disease. "There were love-potions and potions for the preservation of perennial youth, and if treated the right way even gold could be extracted from this miraculous plant." Spaniards converted tobacco to gold.

One wonders why it took so long for people to realize that tobacco was not of any help at all except for the psychological functions that all addictive substances play to improve a patient's psyche, and for the benefit of lining the pockets of entrepreneurs and quacks. Stepheus, the physician of Charles V King of Spain and the Holy Roman Empire (1515–45), ordered cataplasms of fresh leaves of tobacco for the king's goiter, according to Charles Clasius.[9] Monardes, like Nicot, referred to the medicinal properties of tobacco as a "cure it all plant," a panacea. Monardes had been duped by the explorers and physicians coming back from the New World with fantastic tales of cures by the natives. There were no effective medicines to cure illnesses at that time; therefore, there was a genuine desire to come up with better medications. That provided the opportunity for quacks to take advantage of the new herbs brought back from the New World, among them tobacco. The Europeans, being Christian, didn't take into account that the indians practiced "magic" healing.

"This is a herb of Marvelous virtue against wounds, ulcers, noli me tangere (cancer), herpes, and all other things."
Jean Nicot

Tobacco (N. tabacum) was first introduced in France by André Thevet in 1557 when he came back to Angoulem from his trip to Brazil, where he had stayed for three months. In 1559, Jean Nicot of Nimes, (Fig. 45) the French ambassador to Portugal, had as his main mission to negotiate the marriage of the young King Sebastian to the sixteen-year-old Marguerite de Valois, daughter of Henry II.[11] Nicot failed in that mission, but he observed the new tobacco plant (N. rustica) that was smoked by the sailors coming back from the New World. Nicot planted some seeds in the French embassy and soon grew beautiful plants that were shown to one of his good friends, Damião de Goes, (Fig. 46) a Flemish gentleman[12] who gave him information about the uses of this plant that was supposed to cure some cancers.[13] Supposedly de Goes grew this plant in the Royal Gardens of Lisbon, where Nicot had first become acquainted with tobacco.

Nicot decided to obtain more information about the uses of this plant, which was supposed to be effective for the treatment of all diseases including *noli me tangere* (cancer). He wrote to Cardinal Francis of Lorraine of France in 1560 about this wonderful plant brought from the New World, and promised to send a small plant in a pot and seeds to be planted in the French Court. It was not until he went to France himself that he taught the French how to sniff snuff through the nose.[14]

Fig. 45. Jean Nicot of Nimes (1559). Ambassador to Portugal where he was introduced to tobacco by Damião de Goes. He brought tobacco to the French court of Catherine de Medici as a cure for her migraine headaches. Painting by Henry Glatzius. Collection Gorbaty. Berlin 1558. Germany.

There is another version: that Nicot went to visit the prison of the King of Portugal and the keeper of that prison introduced him to the plant (N. rustica) that had been brought from 'Florida'. B. Laufer 146–147.

Lorraine's brother went to Portugal soon after and was shown Nicot's garden with the beautiful tobacco plants in bloom. The seeds and plant were sent as a gift to Catherine de Medici, Queen of France, who was led to believe in the supernatural and medicinal properties of this plant and is supposed to have used it to cure her migraine headaches. The plant took on the name of *L' Herb Médicée*. Catherine and her son Françoi II, became inveterate users of snuff.[15]

Fig. 46. Damião de Goes (1501–1574). Elite at the Portuguese, Court, introduced Nicot to the tobacco plant as a panacea plant to cure all illnesses. Drawing by Albert Dürer (1471–1528.) Albertin Museum.

Soon after tobacco entered France, it was brought to Italy by Nicolo Tornabuoni, agent to Cosimo the Great in Paris, first Grand Duke of Tuscany, who created the first tobacco monopoly in Italy in 1627. Tornabuoni introduced it as a medicinal herb.

Johannes Neander of Bremen eulogized tobacco as a cure-all plant. (Fig. 47)

Fig. 47. Johannes Neander. Eulogized tobacco as a medicinal plant, a panacea. Tobacologia. Leyden 1622

12.

THE GLOBAL LEAF
The Brotherhood of the Leaf

The English were the most influential in propagating the use of tobacco and the pipe during the thirty years war in Europe and then to Eastern Europe and Russia. Tobacco's next stop was Asia, where it first showed up in large amounts about 1575 when Spaniards began trading it in the Philippines. Soon tobacco was being grown in the Philippines. About twenty-five years later, Chinese sailors and merchants introduced tobacco to the southeastern Chinese province of Fukien, from where it spread to the rest of China and then to Tibet and Mongolia. The Portuguese transmitted tobacco to most of the rest of Asia in the first years of the 17th century via Japan and Macao.[16] From there it spread to Korea,[17] India, and Ceylon in the early 1700's. During the same period tobacco was being carried to Persia[18] and the Ottoman Empire.[19] By the third decade of the 17th century, not only the leaf, but the seeds of tobacco had been introduced in large amounts into West Africa. A Dutch traveler in Guinea in the 17th century related that "some Negroes have pipes made of reeds which are about six feet long to the end of which is fixed a stone earthen bowl, so large that they cram in two or three handfuls of tobacco which pipe does filled they without ceasing can easily smoke out, and they are not part to hold their pipe, for . . . it rests on the ground."[20]

The people of the world had created the "Brotherhood of the Leaf," that "silent society" would not be broken up for the next five centuries. (Fig. 48)

Fig. 48. 'Tabaco' (Tobacco) The brotherhood of the leaf. A 17th C. lithograph describing how universal every form of tobacco use had become. Collection Condomex. Jose N. Iturrieya. México.

13.

INTRODUCTION OF TOBACCO TO ENGLAND, HOLLAND, AND THE REST OF THE WORLD

"When all is said and done, I claim:
A pipe is better than a Dame."
Anonymous, 1650

Tobacco was first mentioned in England through the translation by John Frampton of Nicolás Monardes's book on this subject in 1574, *Joyfull Newes oute of the 'Newe Founde Worlde'* (London, 1577). England had been slow in entering the tobacco trade. Portugal and Spain had the largest navies in Europe and they had remained unchallenged for the first half of the 16th century. They maintained a monopoly on all products of the New World—gold, silver, wood, spices, sugar, tobacco, and other products. The Spanish Armada remained unchallenged until after the death of Queen Mary of England, wife of Philip II, who had tried to bring the Catholic religion back to England. Under Queen Elizabeth, (1558–1603) "The Virgin Queen," England became protestant and therefore out of the jurisdiction of the Pope's mandates. Elizabeth realized that the English had to be part of the New World and have to challenge the Spanish Armada, building a strong navy and starting overseas piracy of the Armada in 1562. The peace between England and Spain was over.

Sir Francis Drake in 1562[21] (Fig. 49) began his twenty-five years of piracy that did not end until the Spanish Armada was finally defeated in 1588. By defeating the Armada, England now had access to the markets of the West Indies and was able to exploit the large demand in Europe

Fig. 49. Sir Francis Drake (center), Hawkins (left) and Cavendish (right.) American Heritage 1963

for tobacco and sugar. Drake introduced tobacco to England on a minor scale, first in 1570, and then again in 1573.[22]

The English wanted to be part of the conquest of the new lands, and ventured into territories that were unexplored by Spain and Portugal, primarily in the northern continent. Hawkins described the smoking habits of the Indians on his first exploration of the northeastern corridor of America. While on his way back to England in 1565, Hawkins got in touch with the settlers of the French Huguenot expedition at Fort Carolina.[23] Hawkins had already observed the smoking of tobacco by the Timucua (or Timucuan) Indians, and now he witnessed smoking by the French settlers as well.

Hawkins, a close friend of Sir Walter Raleigh, brought tobacco to England in 1565. He played an important role in the introduction of the smoking of tobacco to England, and also in influencing Raleigh and Queen Elizabeth to send an expedition to settle a colony in the lands that he had explored along the seaboard of North America. Raleigh was no stranger to tobacco. According to MacKenzie, Raleigh's account of his voyage to Guiana during the first Orinoco expeditions (1595) mentions that he encountered "tribes handsome and healthy, who were greatly addicted to tobacco." On his trip back to England from Guiana, Raleigh brought along dry tobacco so he could have it for a long time to come.[24]

Sir Walter Raleigh, the "Admiral of the seas" or "General on land" appointed by the Queen Elizabeth, was given the go-ahead to explore, colonize, and govern that part of the new continent, and he became the founder of Virginia (named after the "Virgin Queen").[25] The first exploration to establish a settlement was undertaken by Raleigh's cousin, R. Greenville (1585), who landed in Albermarle Sound of North Carolina (near Roanoke Island) where he left behind 107 men under Captain Lane and Hariot. They would spend one and a half years there. Sir Francis Drake took the survivors of this expedition back to England in 1586. They were in terrible condition.[26] Hariot, a mathematician and friend of Raleigh, made a study and a survey of those lands, and on his return to England he reported on the common use of pipe smoking by the Indians.

Raleigh, who never set foot in Virginia, soon became addicted to pipe smoking. A story goes that at the end of the 16th century, when pipe smoking was just starting to be the custom in England, Raleigh was sitting comfortably in his chair smoking his pipe one day, lost in his thoughts between large clouds of smoke when "his gardener entered his room and saw large clouds of smoke coming from his master's mouth and in terror brought a bucket of water which he threw on him thinking that his unfortunate master had caught fire and was going to die."[27]

Raleigh recognized the potential value of the cultivation of tobacco, from which large profits could be obtained. Sailors in the ports of England were already smoking tobacco, to the consternation of the populace.[28] Tobacco was smuggled into England from Spain and Holland, creating a large black market. The Dutch had been smuggling it into England as early as 1575. Raleigh was influential in converting many of his friends and acquaintances in the court to pipe smoking.[29] He even taught Queen Elizabeth how to smoke a pipe.[30] Tobacco was valued in England for the pleasure it provided, rather than for its beauty or "medicinal" properties as in Spain.

This habit spread like wildfire throughout all of England. Smoking a pipe (or "Drunk Tobacco," as it was called at that time) became a part of being a gentleman. It was considered to be a disgrace not to be able to smoke properly and to inhale and exhale the smoke aristocratically. There were schools for the well-to-do youngsters to teach them how to smoke properly. "In England 125 pounds were paid for a pound of tobacco. A professor of teaching children how to smoke was allowed to spank them if they forgot their medicinal pipe and they could be punished by going to jail or having to go to work to plant tobacco."[31] There were special places in the cities where people would gather to enjoy smoking (Tabagies), where special chairs and hot cinders to light their pipes were provided.

In 1638 1,400,000 lbs. of tobacco was already being supplied to England from Virginia and in 1775 100,000,000 lbs. of tobacco was imported. Raleigh was named the "patron" of pipe smoking. Virginia would eventually become the greatest tobacco plantation in the New World and be the catalyst for the birth of the most powerful nation on earth. Tobacco was without doubt one of the most attractive and powerful forces for the colonization of this part of America, and the main factor in the slave trade that ensued.

At this point the Europeans were not interested anymore in looking for "El Dorado," as Sir Walter Raleigh had done on his first exploration of Guiana (Orinoco–1595). They wanted tobacco now.

King James I prohibited tobacco use in England at the beginning of his reign in 1603. Two dreams of Sir Walter Raleigh came true—the tobacco plantations in Virginia were a *fait accompli*, and an English colony where English was spoken was well established in America—but only after he had been guillotined by James I in 1618.

Raleigh was a friend of two prominent men of his time—his neighbor, the poet Edmund Spenser (1552–1599), and William Shakespeare. Spenser is most famous for an allegoric poem, *The Fairie Queen*, (1590) which glorified England and the English language. Spenser wrote of "Divine Tobacco" and praised the curative effects of the plant:

> When Belphoebe went hunting, found
> the wounded Knight Timais, one of the
> herbs she gathered in the woods to heal
> him was Divine Tobacco.[33]

Shakespeare did not mention tobacco in any of his writings.[34]

England would have control over the new colonies of America for the next two hundred years, exploiting the lands and its natives and also bringing black slaves to America to work in the sugar and tobacco plantations (and later cotton plantations). There was a need for a large labor force to work on these fast-growing plantations, and the easiest place to obtain slaves was Africa. Thus began the infamous slave triangle that went from England to Africa, to America, and back to England again, with boats loaded with tobacco.

The tobacco/slave boats have been described at the Smithsonian Museum as smelling of tobacco and hemp, but in reality these boats had the stench of the poor slaves that were brought from the coast of Africa. The slaves lived in these boats for long periods of time in cramped, confined, smelly spaces where often they died. The saddest part of the story is that they often were captured and sold by their own brothers.[35] The slaves were often bartered in exchange for tobacco. Most blacks brought to America were first imprisoned on the coast of Africa in forts such as the infamous Dutch Gorée Island Fort,[36] before they were picked up by Spanish, Portuguese, Dutch, and English slave boats.[37] The first boats that brought slaves to North America, in 1619, were of Dutch ownership. The slaves had been bought with tobacco and, after unloading their human cargo, the Dutch filled their boats with tobacco from Virginia.

The number of slaves brought to North America between 1619 to 1680 was about one thousand a year. That number increased considerably from 1680 to 1730, to four to five thousand a year. As the United States grew and became more prosperous, partly because of its tobacco plantations, that number escalated to forty to sixty thousand a year, and then to one hundred thousand a year in the early 1800s.

The first tobacco plantation in England, in 1586, was unsuccessful because of poor financing and poor management, but also because the tobacco that was planted was N. Rustica.[38] This kind of tobacco was fine for the natives of North America, but proved too strong for the English palate. which was already accustomed to and enjoying the sweet Spanish "Orinoco" tobacco (N. Tabacum) from Venezuela (Veriné petum—1639), Hispaniola, and Brazil. This tobacco was "perfumed" and milder. Sir Walter Raleigh, who had made pipe smoking socially acceptable and popular in England, sowed tobacco seeds on his farm at Youghal, Ireland, but without much success. Again, the tobacco Raleigh sowed, N. Rustica, was too dull and earthy, and "nay the Virginia tobacco."

William Barclay a pro-tobacconist from Edinburg in 1614 recommended that only Virginia Sweet Tobacco should be sold in England: "Albeit this herb disdains not to be nourished in many gardens in Spain, in Italie, France, Flanders, Germanie and Brittaine. Yet never the less only that which is fostered in India [America] and brought home by mariners and traffiquers [Spanish and Dutch] is to be used." England was a large market for the "Spanish Leaf," which prompted England to plant its own tobacco in the new settlements of Virginia and Maryland and thus save the expense of buying it from Spain. Increased taxation of Spanish tobacco had not stopped its flow into England and created a black market for smugglers. The time was ripe.

Spain had created a tobacco monopoly in Europe. All tobacco from the Spanish colonies was to be sent to the Port of Seville and shipped in

"Spanish Bottoms" (galleons). Direct sales from the colonies to foreigners were illegal, and violators would be put to death. The result was the creation of a black market. On the other hand, the law benefited Seville as the main manufacturer of cigars and other tobacco items. England soon realized that the "stinky weed" of James I could bring high tax revenue to the coffers of the treasury of England. Furthermore, many merchants would acquire great wealth, and a segment of the population that was in dire poverty would gain employment. The same argument given today by tobacco growers and/or politicians. James I and his son, Charles I, eventually created their own tobacco monopoly and prohibited the planting of tobacco in England to prevent efforts at bypassing taxation.

Finally, in 1611, after previous failures at planting N. Rustica in Virginia, John Rolfe succeeded in importing seeds of N. Tabacum[39] from Trinidad. (Orinoco tobacco) Rolfe's foresight gave new life to the colony of Virginia. A successful plantation was established by 1613, and the first shipment to England of this superior tobacco took place in 1614. (Four years earlier, Rolfe had exported N. Rustica tobacco from Virginia to England, but that crop was considered of poor quality: "A small plant with a little yellow flower liken to Henban . . . whereas the best tobacco of Trinidad and Orinoco is large, sharp and growing two or three yards from the ground, growing like a flower of the breath of our Bell-Flowers in England."[40] Settlers that were ready to return to England in 1606 after three years of near starvation were finally encouraged to stay. They were not interested in cultivating corn that gave only one-sixth of the profit of tobacco.[41]

In 1619 Rolfe recorded, "there came in a Dutch man of warre that sold us twenty negars, they were paid with tobacco." This was the beginning of black slavery in North America.[42] Jamestown was the first enduring and permanent colony of North America, and became the capital of the colony. Jamestown was damaged by fire on May 14, 1607. The capital was moved to Williamsburg in 1699.

"Let England knowe our willingnesse
For that our work is good
We hope to plant a nation
Where none before had stood."
Popular verse, London, 1610

In the June 2002 issue of *The National Geographic*, archeologist Kelso reported on the finding of Ft. James (Jamestown) close to the memorial monument of Pocahontas. Many small clay pipes used by the first colonizers of Virginia were found at that site.

Fig. 50. A Chesapeake Plantation. 17th c. American Colony. Blacks packing tobacco into hogsheads at a Chesapeake plantation, for exporting to England. NYPL

"There was no tobacco that was more pleasant, sweet and strong" than that from Virginia, declared Ralph Hamor, secretary of Virginia.[43] This tobacco would be improved by crossbreeding with other varieties and was then named "Sweet Scented." Some settlers cheated by drenching the English tobacco in a decoction or infusion of Indian Tobacco. Tobacco brought from Virginia to England in 1614 "was sold then for its wayte in silver."

From 1615 to 1616, Virginia was exporting only 2,300 lbs. of tobacco to London versus 58,300 lbs. that came from Spain, but after a slow start in acceptance, the sweet scent of this new tobacco became its best advertisement. Suddenly its demand increased, with England importing all the tobacco that the colony of Virginia could produce, according to Captain Smith, the first governor of Virginia. "All of Virginia went promptly rooting in the ground about tobacco like swine." The first American tobacco enterprise had taken over the colony.[44] By 1617, according to Smith (who disliked tobacco),[45] "every plot from the church down was planted with tobacco."

England began to be a strong competitor with Spain in the tobacco market, not only in the important English market, but also elsewhere in

Europe. In 1617, the colony of Virginia was already exporting 20,000 pounds.[46] In the next ten years they caught up with the Spaniards, and the main tobacco market was now in London. This happened in spite of the initial opposition to smoking by King James I, who eventually succumbed to the charms of tobacco taxes. James used this money to finance the Thirty Years War against France from 1618–1648.

The stage was now set for tobacco to play a leading part in the economies of the seventeenth century and beyond.

Life on the Tobacco Plantation in Virginia and Maryland

"The price of a wife shall have precedent of all other debts in recovery and payment, because this merchandise of all others was deemed the most desirable: 120 lbs of tobacco of the best leaf for a wife."[47]

The Virginia Tobacco Company

On the plantations in the colonies of Virginia and Maryland, tobacco was cultivated by male and female slaves who toiled as field hands planting, topping, suckering, controlling pests, and harvesting. Life for the slaves and their families on these plantations was hard and precarious, but the white middle-managerial employees had a good life, and the owners an opulent one. (Fig. 50) Such was the case on the plantations of George Washington, (Fig. 51) who himself sniffed snuff, as well as on the plantations of Thomas Jefferson, William Byrd and others. Slavery

Fig. 51. Washington sharing the calumet with an Iroquois Chief.

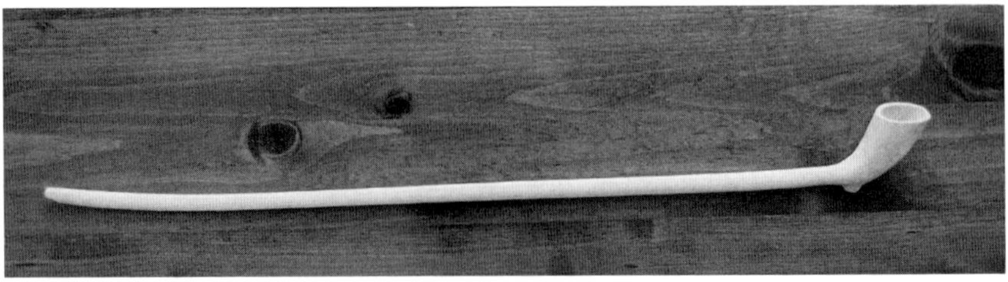

Fig. 52. American Colonial Pipe with long 'hygenic' stem. The end of the stem was broken off every time it was smoked. Tobacco Shop in Ridgewood, New Jersey. Courtesy of the owner. Photography by R. Oropeza.

continued from the 17th century until its abolition during Lincoln's presidency near the end of the Civil War.

The tax on tobacco that was imposed by England on the colonies would be the main trigger for the War of Independence. The clergy (seventy of them in Virginia) demonstrated because they were obligated to take a small price for their tobacco that brought big profits to the crown of England.[48] Defendants appointed a young maverick lawyer, Patrick Henry, to represent them. Henry demanded the right to pay the clergy in cash instead of tobacco. (This followed the 1758 "Parson's Cause" trial, which championed the right of Virginia to pass her own laws such as the Two Penny Act, where debts could be paid in currency rather than tobacco.)[49] Henry assailed the king as "unworthy of obedience no longer a father." One priest was awarded "one penny damages." Patrick Henry went on to immortality for his address to the Virginia Convention of 1775 in which he declared, "Is life so dear, or peace so sweet, as to be purchased at the price of chains and slavery? Forbid it, Almighty God! I know not what course others may take, but as for me, give me liberty or give me death." Long stem pipes were often smoked (Fig 52).

Buying, Selling, and Smuggling Too

The merchants of London seized control of the commercial tobacco trade after 1624, which saw the dissolution of the monopolistic Virginia Company.[50] At that time London was the only port that the crown permitted the Chesapeake tobacco to enter. From there, these small merchants (later, larger companies gained ascendancy) traded and exported tobacco throughout Europe, with Holland becoming the main European market. In the late 17th century, other English ports began to import colonial tobacco. So lucrative were the revenues from tobacco, that royal tobacco monopolies were created in Europe in the second decade of the 17th century. The governments had surrendered to tobacco. Virginia and Maryland alone produced enough tobacco to supply the English and European markets. Of great concern to these interests was how to control, if not prevent, the flourishing contraband trade.

Holland and Tobacco. Introduced to Holland as Hyoscimous Luteus or Yellow Henbane in 1554

"Here one sees tobacco growing on its stalk
The Plant that was brought here from far away places
from the fourth continent of the world
with its double corolla it stands for in all its beauty."

An old Dutch poem, 17th century

In 1590 when smoking was just taking hold in England, Holland was not only getting into the habit as well, it was also involved in growing of tobacco both in Holland and in its colonies on a large scale and smugling tobacco to England in large amounts. Dutch ships transported tobacco to Europe from North America after unloading slaves brought from Africa.

The Dutch colony of New Amsterdam (Manhattan) had been bought for $25 ($2,000 today) from the Canarsie Indians, and it was settled in 1625. From 1637 to 1647, the director general of the colony, William Kieft, ordered a ban on smoking. There were mass protests from the population. Kieft's antismoking policy was followed by his successor, Governor Peter Stuyvesant. The statue of Peter (or Petrus) Stuyvesant can still be seen in the private Gramercy Park of New York City. The smoking ban continued to cause friction not only with the settlers, but also with the Indians. The Dutch were in competition with the English for the tobacco market in Europe, which was one reason why the English wrested New Amsterdam from the Dutch in 1664.

Tobacco was smoked mainly in pipes in New Amsterdam. That custom was also practiced by colonial Dutch women, who entertained each other while idle at home

The Dutch also had settlements in Pernambuco and Recife in Brazil, but they were expelled from there by the Portuguese in 1664. A small group of Jews migrated to the island of Curaçao (North of Venezuela) where they founded the first synagogue in the Americas in 1654. Most of them soon left Curaçao penniless and immigrated to the colony of Amsterdam (New York City). This was the first migration of Jews to America. The Jewish community of New York celebrated this event in September 2004.

There is a saying that in Ceylon, one of the Dutch colonies of Southeast Asia, "men begin their day with gin and tobacco and end it with tobacco and gin."

The Dutch improved not only the cultivation, but also the curing of tobacco and its manufacturing.

Tobacco for Medicinal Purposes in Holland and England

> "Singing and Playing Joyful Farmers"
> Oh fragrant smoke, which
> vies with violets and sweet
> roses in smell you have
> stolen my heart and my
> fancy although your color
> is very miserable.
> You defy the best medicine
> power, with your perfume,
> you restored me, when I
> was pining away.
> From the song "Joaghim," 1716
> Words by Jan Van Gigsen, Music by A.Van Hulkenroy

Europeans were aware of the use of tobacco by the American Indians as a ceremonial ritual, as a stimulant substance, and for medicinal use.

Tobacco in the 16th century was first found in Holland in botanical gardens, but then it was planted in a large scale as early as 1615. Tobacco was used as a cure-all (panacea) for many illnesses. The famous Jan Van Beverwijk (1594–1647), first parish doctor and professor at Ultrech, wrote in his Wealth of Health:

"It [tobacco] has a wonderful way of ridding the head of all moisture and phlegm. It has the tendency to promote desiccation. It turns the stomach and loosens the bowels and causes violent convulsions throughout the body . . . It is an excellent remedy against the plague and reliable persons have told me, that when a great many people in London died of the plague, those who frequented the tobacco houses, never caught the disease."[51]

Tobacco in 16th-century Holland was mainly processed for medicinal purposes. The leaves were boiled with vinegar or wine, and evaporated and mixed with anis seed and ginger. Apothecaries crushed the leaves of tobacco in a mortar to make snuff for medical purposes.[52]

One Dutch prescription for tobacco dictated, "Tobacco should be taken early in the morning, on an empty stomach . . . not by children or by those who are mentally deficient or by those who have high blood pressure; but by those who are strong, cold and have low blood pressure and often suffer from catarrh."[53] Tobacco was recommended for all the ills to which the flesh is susceptible, but for syphilis J. Neander (1622) recommended that "it would be better to abstain from visiting brothels."[54]

Students of the University of Leyden came from all countries in Europe to receive higher education, and they were the best ambassadors for bringing tobacco home to their own countries. Dr. William Van der Meer, physician at Delft, cultivated tobacco in his garden. Seventeenth-century physician Dr. Johannes Vittich mentioned that during the plague of 1664–1666, even children were made to smoke tobacco each morning.

Russia, like Holland (and every other country), agreed with the medicinal properties of tobacco. Tobacco, it was said, had "Marvelous Healing Virtues."[55] The Russians agreed with other Europeans and the natives of America that "There can be no doubt that tobacco can cleanse all impurities and disperse every gross and viscous humor, as we find by daily experience. It cures cancer of the breast, open and eating sores, scabs and scratches, however poisonous and septic, goiter, broken limbs, erysipelas and many other things."

By the middle of the 19th century, tobacco had almost disappeared from the Materia Medica in the Pharmacopeias; however, up to the present time one will still find tobacco as one of the folk remedies specifically used by quacks or *curanderos* (shamans) in Latin America, *santería* in Cuba, or voodoo in the Dominican Republic and Haiti.

There were non-believers in this cure-all plant. King James I of England sarcastically wrote at the beginning of the 16th century: "O Sovereign Tobacco! That art of medicine for every Malady, a salve fore every sore. 'Twill cure the dropsy, the gout, the rheum, the cold, the ache of the head . . . It will make a woman that is barren to bear six children in one night. Will make a lean man fat, and a fat man lean. But I know it hath made wise men to become fools."[56]

Tobacco in the Rest of Europe and Elsewhere

War in Europe increased the use of tobacco. During the Thirty Years War of the 17th century, the use of tobacco spread eastward to Bohemia,

Austria, Hungary, and the rest of Europe. One reason was that most of Europe was still under the influence of the Habsburgs. One of the main

Fig. 53. Soldier of the Thirty Years War 1618–1648 smoking his pipe in a moment of rest. The pipe became a great friend of the solitary soldier. It soothed his nerves and consoled him while he was under stress or idle. By John Toorenuliet. Photo Berlin

pastimes when soldiers were stressed and idle was to smoke their pipes. (Fig. 53) They would pass the habit on to friends and foe because of the close contact they had with soldiers from other nations. In this way Germany, Austria, Holland, and England (Fig. 54)[57] soon took up the habit. "Is he moody, angry, or perplexed, he sticks his pipe between his teeth and takes a long pull at it. Should his wife begin to nag, the man will fill his mouth with smoke and puff it in her face."[58]

The Swedes were introduced to tobacco in 1616. They smoked it and chewed it. The Italians, still part of the Holy Roman Empire in the 17th century, eagerly took up this practice through the influence of the sailors coming into the ports of Naples and Sicily from Portugal and Spain.

The Introduction of Tobacco to Turkey

Tobacco was introduced to Turkey at the beginning of the 16th century. While on a trip on the Dardanelles in 1599, John Dallam, the organ

Fig. 54. In the time of the Stuarts, both men and women frequented the English tavern (Tabagie) to socialize, drink ale and smoke. Smoking tobacco was a favorite diversion of both the English and the Dutch. From the Satite of John Suckin. British Museum

builder, wrote in his book *Early Voyages of the Levant* that a Turkish captain requested a gift of tobacco and a pipe while boarding the ship for inspection, which illustrates that the Turks were familiar with its use. By 1615 the Turks had taken to smoking with gusto. Thevenot and Pietro-Valle wrote that the Turks were pipe smokers, like the English and Dutch: "The Smoking of Tobacco is committed to such an extent by the Turks, that they are rarely to be seen without smoking." They stated that this incessant smoking gave the Turks an advantage over the Christians because they had time for reflection before given an answer to a transaction.[59] This happened before the Thirty Years War spread tobacco use throughout Europe.

The Turks grew tobacco mainly in Macedonia; now it is cultivated in many parts of Turkey.[60] They smoked (and still do) a stronger tobacco, n. tumbak (n. alata). (Fig. C32) The Armenian and Jewish population in Turkey consumed twice as much tobacco as the rest of the population.

From Central Europe, this custom was brought to Russia by land and through the Baltic. Poland acquired the use of tobacco not only through direct contact with other European countries, but also through a "valuable" present from Sultan Murad III (1623–40). "Some dried leaves of the precious novelty" (tobacco) was considered a unique present to the Polish King. Murad III (and later his son, Murad IV) issued severe punishments, including the death penalty, to any Muslim who was found smoking, but apparently he found it politically correct to give tobacco to the King of Poland.[61]

While in Hamburg, Germany, in 1798–1799, Samuel Taylor Coleridge found Germans "smoking pipes of all shapes and fancies . . . cane, clay, porcelain, wood, tin, silver and ivory. Many with silver chains and silver covers." He warned a fellow that was about to join him in Germany "to be prepared to put up with German Smoke," and he noticed that "male teeth were commonly black and yellow with life's smoking." (Today we see this dark brown discoloration of the teeth and fingers in Thailand and Burma, where they chew betel nut mixed with lime and tobacco.)

Japan was the first Asian country to smoke tobacco on a large scale. It had been introduced by Portuguese onboard a Chinese pirate ship in 1542 that was forced to land on the south shore of Japan during a storm. The Portuguese returned in 1595, offering tobacco to the inhabitants of Kyushu and teaching them how to smoke.[62] The Shimatsu prince of the House of Satsuma received tobacco leaves and seeds that he sent to the Emperor of Kyoto (then the capital of Japan), who planted them in the Royal Gardens.

By 1603 smoking was endemic in Japanese society, and still is. Like the English, they established "smoking clubs" or tabagies. The Japanese wrapped their "Tabako" in white rice "Hoshu paper." They also smoked tobacco in small pipes of palm wood with long stems and silver mouthpieces, taking only a few small puffs from it before they had to repack it and relight it. These pipes were called kisseros. Tobacco was even in use on the distant Ryukyu Islands in northern Japan. Their tobacco was strong and they often chewed it rather than smoked it.

The Japanese made efforts to stem the use of tobacco, imposing heavy fines and imprisonment; but to no avail. The Japanese took to this bad habit with all the fervor and devotion of their European counterparts. At the end, even the royal family was smoking.

From Japan, tobacco went to Korea where it was known as *nampankay* or *tañbako*.[63] (Korea had been defeated by the Japanese.) And then tobacco took China by storm during the Ming Dynasty.[64] From there it invaded Manchuria and Tibet. The Chinese expressed words of endearment about tobacco such as "smoke blossom," "herb of amiability," and "herb of discernment."[65] Tobacco had already been introduced to Macao in 1567 by the Portuguese,[66] and at the beginning of the 17th century there was a heavy trade in the weed between the Philippines and Canton China. The Spaniards built a huge cigar factory in Manila, a "copy cat" version of the one in Seville, where hundreds of women worked.

Tobacco in Africa and India

Beginning in the 16th century, wherever the Spanish, Portuguese, or Dutch mariners landed along the coasts of Europe, Africa, or Asia, they used the charms of tobacco as a "softener" while buying other commodities such as silk, china, spices or other items. They used tobacco for barter, gifts, or bribes whenever needed.[67] A Dutch resident in Guinea (1595) wrote that "some negroes have pipes made of reeds, which are about six feet long, to the end of which is fixed a stone or earthen bowl, so large that they cram in two or three handfuls of tobacco, which Pipe thus filled they without ceasing can easily smoke out, and they are not put to hold their pipe, for . . . it rests on the Ground."[68]

From the African ports, tobacco penetrated inland by caravans to distant sites. Tobacco spread to North Africa from Egypt through the Sahara and the Nile River.[69] When the explorer Mango Park went through West Africa's Niger River in 1795, he recorded that all the natives took snuff and smoked tobacco. The pipe bowl was sometimes made of clay or wood with a stem of iron shielded with ivory. (Fig. C33) While looking for the origin of the Nile, Livingston narrated that the villagers of Luanda, both men and women, smoked tobacco in long pipes.

The Africans started their own plantations once they became addicted to tobacco, but they preferred to buy the already cured leaves from the Portuguese and the Dutch because it was ready to be smoked. The

Portuguese commented on "The greedy appetite for smoke among these negro people." The Portuguese planted tobacco in the Cape Verde Islands where they had built a fort to hold slaves. Slaves were sometimes paid for with tobacco, just like in North America. The demand for tobacco in Europe and Africa fueled the demand for a large labor force to plant tobacco, thus increasing the trade in black slavery. Other commodities that were in demand were horses, silk, iron spears, and firearms.

Australia didn't receive tobacco until the 17th century, but the aborigenes chewed the leaves of Nicotiana Suavolens prior to contact with the white man. B. Laufer p. 481

By the end of the 17th century, smoking and sniffing tobacco was widespread in India.[70] Tobacco had been brought to India by the Portuguese in 1530 and had spread through the whole continent in only twenty years. The Portuguese had established a colony in Goa on the west coast of India. India developed her own water pipe, the hookah. They also chewed tobacco with betel nut and lime. India today has a high incidence of cancer of the mouth.

Robert Louis Stevenson (1850–1894) the Edinburgh author of *Treasure Island* and other books was well-traveled. He mentioned that on the island of Paumotu (Samoa) in the South Pacific, "the men spent their time cleaning their vessels with their pipes lighted even when the pipe was only sticking out of the water while submerged."

By the end of the 17th century, the word "tobacco" had been included in all languages in every corner of the world.

14.

ANTI-TOBACCONISTS

"Tobacco in short had a
certain venomous faculty
joined with the heat
thereof."
King James I (1604)

King James I (Fig. 55) and Other Antitobacconists

Until researchers in the 20th century established the causative relationship between tobacco and lung cancer, as well as other ailments such as cardiovascular diseases, most attacks against tobacco through the centuries had been based on a personal dislike of the "unclean" habit, or because of moral, religious, or social biases.

Most of the world embraced tobacco, but the leaf had detractors almost from the beginning of its introduction to Europe. One of the most famous early critics was King James I of England, who in his famous and anonymously published "Counterblast to Tobacco" (1604) decried it as a "custom loathsome to the eye, hateful to the nose, harmful to the brain, and dangerous to the lungs . . . and in the black stinking fume thereof, nearest resembling the horrible stygian smoke of the pit that is bottomless."

James's dislike of tobacco was based on a personal aversion. He referred to smoking as a "filthy practice." James had noticed that "autopsies performed in smokers revealed that their inward parts were soiled and infected with an oily kind of soot." James also disliked tobacco because he was a staunch enemy of Sir Walter Raleigh, (Fig. 56) who was called "the patron of pipe smoking." Raleigh had tried to dethrone James,

Fig. 55. King James I of England. (1603–25) Tobacco was considered by him to be a filthy weed and detrimental to health but not to taxes! Collection of E.L. Gorbaty. Berlin.

and disobeyed his order not to interfere with the Spanish settlement in Orinoco. Raleigh attacked that settlement and was put to the guillotine by James in 1618.[71] William Barclay from Edinburg 1614 was a pro-tobacconist.

C19. Indigenous people of Ahuacatlán, Zacatlán Puebla, México, selling home-grown tobacco in an open market (Tianguiz) as their predecessors did for centuries. (l.) Husband holding a large leaf of tobacco. (r.) Wife weighing tobacco. The church of Ahuacatlán can be seen in the background. Photography by Felipe Álvarez Ricaño. Montage by R. Oropeza.

C20. Angular ceramic pipe. Mexica Culture. Museum of Anthropology of México City. Photography by R. Oropeza.

C21. Ornate angular pipe of San Miguel de Allende Guanajuato, México. Museum of Anthropology of México City. Photography by R. Oropeza.

C22. (left) Fray Bernardino de Sahagún (1503?–1590 Learned Nahuatl, the language of the Mexica (Aztec) as well as their customs, culture and history, and their view of the conquest of México. He published his findings, known as *The Florentine Codex,* in about 1580. (below) Mexica culture pipe carved in the shape of the head of an Aztec. (bottom) Tarasca pipe. Both from the Museum of Anthropology of México City. Photography by R. Oropeza.

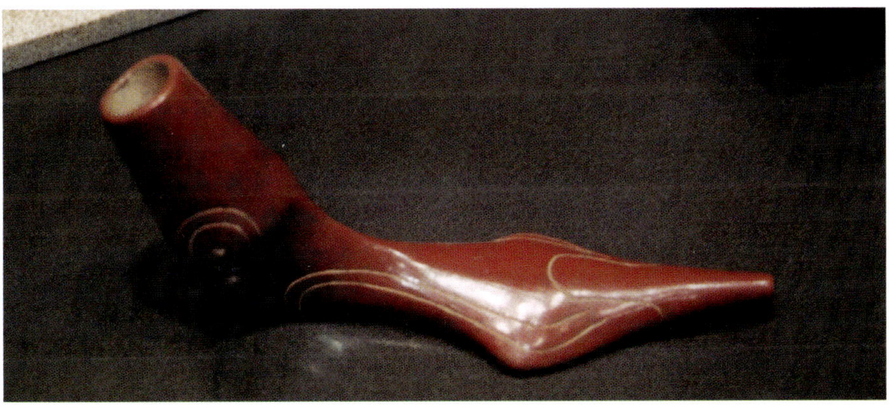

C23. (left) Gifts of Hospitality, sharing lighted tobacco reeds before a civil ceremony. Florentine Codex (Sahagún) 16th c. Photography by R. Oropeza.

C24. (below) Elegant Tarascan pipes from the Museum of Anthropology of Morelia, Michoacán, México. Photography by R. Oropeza.

C25. Typical Tarascan pipe with a high bowl (hornillo) and a long stem. Private collection, Rufino Tamayo. Oaxaca, Oaxaca, México. Photography by R. Oropeza.

C26. Priest of occidental culture (Nayarit) smoking a large pipe. Museum of Anthropology, México City. Photography by R. Oropeza.

C27. (below) Mixtec Mourner (Oaxaca) smoking a ceremonial pipe with a skull as a bowl. Codex Nutall. A gold pectoral from Montealbán Oaxaca is superimposed on the pipe smoker. (Private collection, R.Oropeza)

C28. (left) Chieftain Powhatan of the Chesapeakes, father of Pocahontas, smoking a pipe while Captain Smith was delivered to him in 1607. He protected the settlers during the starvation period. Engraving by William Hole.

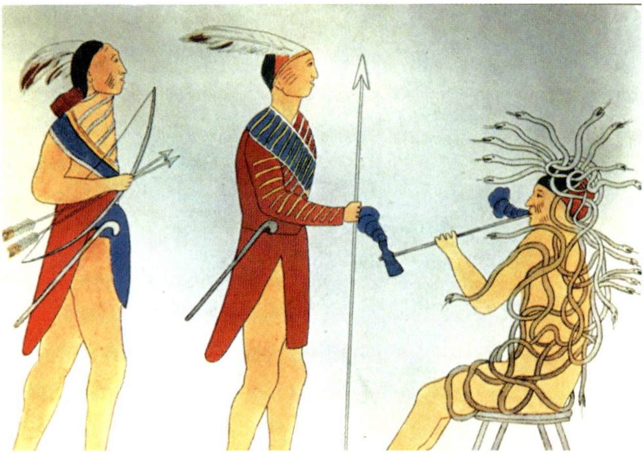

C29. Chief Medusa, Atotarho of the Onondaga of the Iroquois. He consummated the Iroquois confederacy. He smokes his calumet pipe while meeting his British Allies (left). Picture by Seth Eastman.

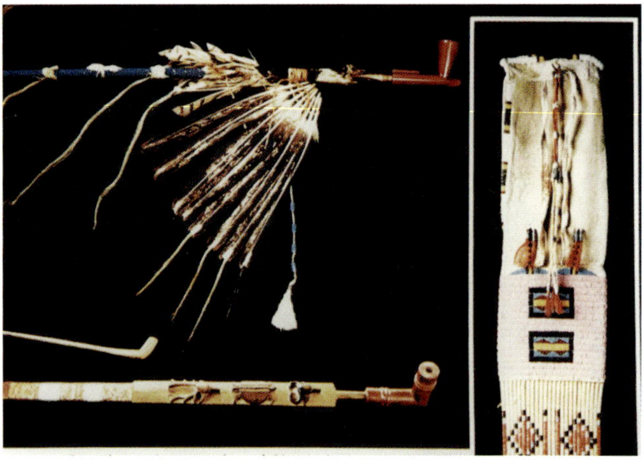

C30. The Calumet Pipe, spiritual pipe of peace or war, was an ornamental pipe with feathers, beads and horsehair from the defeated enemy. "The pipe is us." said one Sioux. "The stem is our backbone. The stone is our blood as red as our skin." Americas' Fascinating Indian Heritage. James A Maxwell.

C31. Map of Missouri Expedition from Ohio to Oregon. Lewis and Clark (1803–1805).

C32. Turk from Istanbul puffing on his cigarette of strong nicotiana. Photography by R. Oropeza. 1999.

C33. African pipe with a bowl made of clay. The stem is made of iron shielded with ivory. Collection, R. Oropeza.

C34. The Hookah. 19th c. Made in Austria by The Swarovski Crystal Company. Collection R. Oropeza.

C35. Carving a Meerschaum pipe for the author. Capadoccia, Turkey. Photography by R. Oropeza. 1999.

C36. Gentlemen of importance. Mexican aristocrats at the end of the 19th c. Photography from a 19th c magazine.

C37. Emiliano Zapata sharing a cigar with the American Consul, Mr. Carothers, during a meeting in Xochimilco, México, in 1914. They had nothing in common except for the cigar. Capitalism vs. Agrarian Reform. Photography by Agustin Victor Casasola. (INAH) [687563]

C38. La Dolce Vita. Two graceful and beautiful smokers enjoying the waterfalls at La Fontana de Trevi, Rome Italy. Photography by R. Oropeza. 2004.

C39. Scutari (Üskudar) Hospital (now The German Hospital) across the Bosphorus Strait from Istanbul,Turkey. This hospital was used during the Crimean War. The war was a catalyst for the spread of cigarette smoking. Florence Nightingale became famous in this hospital. Photography by R. Oropeza (1999).

Fig. 56. Sir Walter Raleigh (1552–1618) Admiral of the sea and a close friend of Queen Elizabeth I. He urged exploration of the Orinoco in South America and colonization of North America. He brought both tobacco and curare from the Orinoco region. But he was an enemy of King James I who had him guillotined in 1618. From a portrait by an unknown artist in the National Portrait Gallery

King James's brother-in-law, Christian IV of Denmark, claimed that tobacco drove people insane and clogged their brains as well as their lungs. Other detractors considered tobacco one of Satan's most crafty devices.

When King James I died in 1625, his son, Charles I (1625–49), who was also opposed to tobacco but not to the taxes it raised and the profits from his father's newly created monopoly, confirmed the latest proclamation against smoking by his father, this proclamation was not effective. Charles reinforced the monopoly on tobacco in order to continue to control its taxation and profits.[72] In the English colonies of the New World, tobacco was often used in the 17th century as currency because of its high demand and value on the European markets.

Another powerful king of the 17th century, Louis XIV of France, also abhorred tobacco but not taxation of tobacco.[73] Members of his court were pipe smokers, and it was said that when Louis XIV was about to come to court, they all hid their pipes and tobacco in their sleeves. As soon as he left, the pipes immediately appeared lighted in their mouths again as if by magic. Louis XIV's daughter was a heavy pipe smoker. One day she was warned by the Imperial Swiss Guards that the king was on his way, but she only had time to hide the pipe, and not the smoke. The next day she was harshly reprimanded.

Louis XIV, however, allowed the Marquis de Prade to publish a book praising tobacco use. The book was titled *Privilege du Roi* (*Privilege of the King*). Louis also, like James I and other sovereigns of the 17th century, found it financially convenient to tax tobacco and create a monopoly that enriched his coffers. He then looked the other way when not only the court, but the rest of France engaged in pipe smoking. During the ensuing four centuries, all governments of the world followed the example of these kings of the 17th century. Even the Catholic Church "surrendered" to tobacco at the end of the 18th century.

In the 19th century Queen Victoria found smoking as evil and repugnant as her predecessor, King James I, and prohibited its use on the royal premises.[74] Victoria encouraged the population to give up the habit of smoking, and even stated that Adam's fall was likely due to the tobacco pipe! On the other hand, her son, Edward VII, Prince of Wales, was an ardent smoker. When he became king, he reversed the prohibition against tobacco. He would announce, "Gentlemen, you may smoke," when he entered the Coffee Room at the Court. "This marked the end of an era and the beginning of his tenure as a King."[75]

The Spanish writer Francisco de Quevedo at the time of the Catholic Reformation in the 16th century felt that the Indians "have avenged [themselves] by introducing tobacco to their Spanish colonizers and that the tobacco users were as bad as Lutherans." He referred to America as "a rich and beautiful whore."[76]

The Condemnation of Tobacco by the Church (16th–17th century)

"And when the pipe grows foul within,
Think on thy soul defiled with sin."
J. H. Dixon (London)

Tobacco was a commodity for Europeans, something that would give them pleasure or relax them. Tobacco played no part in European religion, but because smoking was part of a religious ritual by the "Barbaric Taíno Tribes of Hispaniola," the American christian church of the 16th-17th century was strongly opposed to tobacco and believed it was the work of the devil. One of the first persecutions for smoking in places of worship was in México in 1575.[77]

Spaniards burned the Native American temples and codices,[78] persecuting, torturing and killing those people who refused conversion to Christianity. These were among the first persecutions of the Inquisition with the approval of the Christian church and the Roman Pope. The first auto-da-fé was in Yucatán in 1535.

Some claimed that tobacco was sold by "Marranos [Jews] and that they employed it against women and children." Jews had just been expelled from Spain in 1492, the same year as the discovery of the New World and the same year as the fall of Granada when the Moors were expelled from Spain. Spain had accomplished her "ethnic cleansing" at last, after ten centuries of occupation by the Moors of Africa.

The colonial priests, by mandate of the Pope and the mighty Spanish church in Hispaniola, were supposed to defend the church's teachings for the good of society, the religious order, and the morals of the colonizers and the natives. They were charged to accomplish this either by the teachings of the scriptures or the edge of the sword.

Excommunication was the penalty for priests who smoked tobacco. Excommunication at that time was a terrible fate for Catholic Christians, who believed that God spoke through the church.

The Councilio Provincial Limense (Provincial Council of Lima, Perú) was held by an archbishop and five bishops in 1583, and its rulings were approved in Rome by the Pope and the Council of Trent on October 26, 1586. The council declared that "under pain of mortal sin . . . and everlasting damnation, no priest, before celebrating and administering communion, could take tobacco in smoke or in any other form [snuff or chewing]. This mandate also applied to the use of coca, which was common among the natives of Perú and Bolivia before the conquest of Pizarro.

In 1585 a similar edict was passed by the church in México[79]: "out of regard for the reverence due to the Holy Eucharist it is commanded that no tobacco in any form whatever be taken by clerics before reciting the holy mass or by any person taking communion." Presiding over this edict was Archbishop Pedro Moya de Contreras, later to become president of the Royal Council of Indies. There were six bishops participating, and this edict also was confirmed by the Pope and approved with the Royal Seal of Spain. These edicts later also became effective in Santo Domingo, Hispaniola, and Manila, Philippines."

Clerical life in the 17th century

"To some clerics:
who smoked very long pipes
and preached very short sermons"

Many priests ignored the warnings by the church, discovered the pleasure of smoking, and became addicted to tobacco. "In some cases they would fall into a state of ecstasy while smoking and would claim that smoking heightened their spiritual vision."[80] Some priests rebelled against the edict because they could not find this mandate in any part of the scriptures. Others would claim that God made everything; therefore, tobacco could not be condemned. Still others argued that smoking would help them abstain from carnal sin.

Prohibitions by the Church in Europe

By 1660 the Protestant Anglican Church of England separated from the Papacy. The Anglican Church did not oppose tobacco, and smoking became rampant in England.

Pope Urban VIII (1623–1644) (Fig. 57) attempted to put a stop to the scandalous use of tobacco by the clergy with a Papal Bull, "Ad futuram rei memoriam," on January 30, 1642.[81] The use of tobacco by the priests of the Church of Seville was an example of what he was trying to stop. Chaplains, priests and deacons either smoked or took snuff during the celebration of mass. Spitting while in the pulpit was common, and some of the priests became so experienced that they spat with precision to the astonishment of the congregation. "Whereas information has lately reached us from the Dean and Chapter of the Metropolitan Church in Seville, that in those parts the use of the herb commonly called Tobacco has gained so strong a hold on persons of both sexes, yea even Priests and Clerks—We blush to State."

This bull was disobeyed, but was reiterated in 1650 by Pope Urban VIII's successor, Pope Innocent X,[82] who prohibited the use of tobacco in St. Peter's Church where "Priests and even the faithful had even the irreverence of spitting on the floor of the church." "One Sunday a priest who was celebrating a mass took a quick pinch of snuff just after receiving

Fig. 57. Pope Urban VIII. Excommunication was used by him as the ultimate punishment for heretics (Galileo Galilei) and tobacco users. Tobacco was against the teachings of the Catholic Church. Maffeo Barbarini. Historia Photo. Berlin, Germany.

holy communion, this was followed by violent sneezing that made him vomit, throwing up the blessed sacrament on the altar in front of the faithful congregation."

Another attempt by the church to stop tobacco use was Pope Benedict XIII's Papal Bull of 1725: "The clergy of the aforesaid cathedral, earnestly, in the name of the Lord, that not only were they to take snuff within the precincts of the church so discretely that no scandal or offense might be given to any of the congregation, but no one, especially when taking part in the holy ceremonies, should openly or covertly offer the box or casket wherein he keeps the tobacco dust [snuff] to any besides or around him."

After the Peace of Westphalia (1648) that concluded the Thirty Years War, another edict against the use of tobacco was issued, this time by Emperor Ferdinand III. At that time the use of tobacco had been widely adopted throughout Europe and beyond. Ferdinand's edict, issued on August 16, 1649, stated in part:

"W.E. Ferdinand, by God's grace Archbishop of Cologne and Elector, do all and each to wit; whereas it is shown; by daily experience that tobacco drinking [smoking] that in former years was used hardly anywhere has now spread abroad everywhere and is so mightily increasing not only among soldiers and common men, but even among children . . . - neighboring princes and states have been compelled to forbid such abuse. . . . Our subjects in general, strictly to forbid the sale and purchase and use of tobacco everywhere under penalty of incurring our high displeasure and punishment, together with the confiscation of the tobacco and the pipes."

The penalties were a fine of fifty gulden or imprisonment, with heavier penalties for repeat offenders. In spite of this edict, however, people kept puffing on the filthy weed.

A manifesto published on May 19, 1653, by ecclesiastics and university professors in Saxony against smoking stated, "It is both Godless and unseemly that the mouth of man, which is the means of entrance and exit of the immortal soul, that mouth which is intended to breathe in the fresh air and to utter the praises of the Most High should be defiled by the indrawing and expelling of tobacco smoke." Again, this edict was completely ignored.

The Electress Maria Anna of Bavaria also prohibited "tobacco drinking in any form," especially by the peasants and common people.[83] This 17th-century prohibition was also ignored. Frederick William II reinforced this law in his Berlin Report 1809. "It is necessary to aim at suppressing this unseemly habit," he declared. This law was ignored when Germany went to war with the France in 1813. In that same year there was an epidemic of cholera, and smoking was not considered harmful anymore. In fact it was considered beneficial and was encouraged. In 1831 the ban against smoking was suspended by the Police Minutoli in Berlin on the basis that smoking was not considered to be a fire hazard in Berlin, as had been declared before.

Prussia's Frederick the Great, a snuffer himself,[84] issued an edict in 1764 that provided stern penalties for smokers. These penalties were even more severe when smoking was done in dangerous places such as wooden buildings. Informants were rewarded for information.

By the end of the 18th century, there was no criticism from the clerics about using tobacco in any form. They considered smoking "a recreation of the traveled spirit." Tobacco, the divine nicotiana, had withstood the first onslaught against its reputation.

Attempts to Ban Smoking in the Turkish Empire

The Mufti clerics maintained that tobacco use was against the teachings of the Koran.[85] They made an analogy between the sinful taking of alcohol and the taking of tobacco. The Turkish poets defied these prohibitions and were also in favor of other stimulants besides tobacco—coffee, opium, and wine. They poetically rationalized these four habits as the "four elements of the world of pleasure, the four cushions of the divan of delight."

After the death of Suleiman the Magnificent, smoking was condemned in the Ottoman Empire by Ahmed I (the same Ahmed I that had given a tobacco plant as a "unique" present to the King of Poland). Those who were found smoking had the stem of their tobacco pipe thrust through their noses, and were paraded on donkeys through the streets of Constantinople (Instanbul).

Ahmed I's son, Murad IV ("Murad The Cruel," 1623–1640), (Fig. 58) who became a bloodthirsty tyrant, also prohibited smoking. Chalebi writes that Murad, "went hard for the smokers Tooth and Nail and tried to close the coffee shops where they smoked and drank, he persecuted them merciless but at the end without success."

Smokers were ready to die for their addictive nicotine, rather than obey the imposed religious beliefs of the Muftis. As the severity of the punishments increased, so did the desire to smoke. Murad IV had punished those who indulged in drinking wine or any other alcohol beverage with the death penalty (though it has been said that he himself drank to excess), and he applied these laws to tobacco users. His orders were carried out by his agents with obedience and without mercy, but people kept smoking in defiance. Murad IV commanded his subordinates to enforce the law. "This is my high behest, which if thou failest to fulfill, thy head shall be cutoff." Murad put to death fifteen or twenty officers who were caught defiantly smoking. Some officers went to their executions hiding their small pipes in their pockets in a final show of defiance, and maybe to have their last wish—a final puff of the sinful weed.

In spite of the horrors of this persecution, the passion for smoking persisted and increased, just as it did in the Christian lands. These ineffectual laws were repealed in 1683 after the Turks were defeated at the Siege

Fig. 58. Sultan Murad IV (1623–1640) According to him alcohol and tobacco use were against the teachings of the Koran, and users had to be dealt with 'tooth and nail.' Death was the ultimate punishment, but the more severe the threat, the more they smoked. Engraving at Vienna National Library

of Vienna. It was whispered that Mohammed IV himself was a smoker. Thereafter, tobacco cultivation became a very important industry and a good source of revenue in the Ottoman Empire. It is still heavily cultivated in Turkey, the last capital of the Ottoman Empire. The Muftis had lost the war gainst tobacco.

Smoking Prohibition in Switzerland (1667)

Berne was the last city to maintain prohibitions against smoking tobacco. The reason was to prevent fires in wooden houses. There were many wooden homes throughout Europe and Russia. Fires were very common and often destroyed large areas of the cities. Some of these fires were smoking related, but this was only an excuse to curb smoking. A fine of 15 liures was imposed on each offense; however, these laws were defied by tobacco planters on the basis that the laws were detrimental to their finances (sounds familiar in the 21st century?) This prohibition, like all others, didn't work.

In other Swiss towns such as Kysburg (1674), authorities stated that the habit of smoking had grown to such an extent that servants and even wives and daughters used tobacco to a scandalous degree and put the town in danger of fire. Some authorities at that time considered smoking to be destructive to health, besides being liable to cause fires, but there was no factual proof to back their opinions.

Russia Tries to Ban Tobacco

Russians had long been addicted to tobacco smoking. Paupers spent their last pennies on tobacco rather than on food.[86] Michael Feodorovitch Romanoff, the first Russian Tsar after Moscow was freed from Polish rule in 1613, was a bitter opponent of smoking. He issued an edict against smoking in 1634 that criminalized tobacco and exiled offenders to Siberia. The patriarchate of the Christian church had convinced him that smoking was not only profane, but also an abominable practice and a deadly sin. Smoking had been declared a deadly sin in 1630 by the Christian clerics in Russia, based on their own interpretation of the Bible passage, "The things that come out of him, those are they that defile the man."[87] (Mark)

Alexis Romanoff, the second tsar, reinforced the tobacco ban through a *ukase* (mandate). The penalties for violating the *ukase* were torture and exile. The Tsar had vigilantes to check for tobacco contraband. When a tavern keeper, tobacco seller, or anyone else in the possession of tobacco was apprehended, they would offer the inspectors a bribe, large or small, from five to thirty rubles, so as not to be taken to the "New Quarter" where criminals were racked and tortured in public, receiving from twenty to twenty-five lashes. Repeat offenders were whipped with such force that they were often whipped to death. The inspector would take the accused's money, but if after receiving it they let their prisoner escape, they could be punished with the knout and expelled from office, according to the civil code of the Tsar. The knout was "a long stout whip applied with force on the back of men or women convicted of using or selling tobacco. It was made of untamed elk-skin that cut like a knife, short as a finger length."[88]

During the reign of Peter the Great (1689–1725), smoking became acceptable. Peter took up the habit of smoking while in Europe, and allowed the importation (and taxation) of tobacco to enhance the coffers of his government.[89]

J. Cruell wrote in the "Present State of Muscovy in 1698":

> Formerly tobacco was so extravagrantly taken, 'as the aqua vitae' and was the cause of frequent mischief: for as much as not only the poorer sort, would rather lay out their money upon tobacco than bread, but also when drunk, so did set their houses on fire through their negligence . . . they take it after a most beastly manner, in the middle of which, there is a hole, where they place the vessel that holds the tobacco. The vessel is commonly made of wood, pretty wide, and indifferently deep; which when they have filled with tobacco, they put water into the horn to temper the smoke . . . they fall down and drunk . . . then they leap up in an instant more brisk and lively than before.[90]

Japan's Attempts to Stop Smoking

Both tobacco and Christianity were forbidden in Japan by the conservative, isolationist Shoganate. Christianity was stopped, but the use of tobacco was not.[91] The first edict against smoking in Japan was in 1603,[92] but was not reinforced until 1607–1609, when the cultivation of tobacco was made a penal offense. In 1612 the Jeyasu Decree was passed, which provided that "The property of any man selling tobacco should be handed to his accuser . . . and anyone arresting a man conveying tobacco on a pack-horse might take both horse and tobacco for his own." In spite of this measure, smoking was prevalent in Japan, even in the Shogun at Yaddo (now Tokyo).

Smoking Bans in the United States

Long before the prohibition of alcohol in 1927, smoking in the streets of Boston was prohibited in 1880. In the year 1927, during prohibition when alcohol was illegal throughout the United States, there were also laws against smoking in twelve states. These laws, like the prohibition against alcohol, were bypassed. Organized crime loved these prohibitions, which encouraged smuggling and created a large black market similar to that in England during the 17th century when tobacco was smuggled in from Spain. The prohibitions soon had to be abandoned. People found ways to get their beloved nicotiana.

In the United States, like everywhere else, tobacco has had its detractors as well as its proponents. Henry Ford hated smoking and insisted that his workers not be smokers, but this ban could never be enforced. Ford had to make peace with the "little white devil," or the "little white imp," as he called tobacco, for the rest of his life.[93]

There were anticigarette advocates in the United States at the end of the 19th century and the beginning of the 20th, among them Lucy Gaston Page.[94] The Women's Christian Temperance Union paraded with placards showing a mother with a cigarette in her mouth and a child on her lap, hoping to deter smoking. These movements were ineffective; people ignored them and went on smoking and drinking. (Fig. 59)

Fig. 59. Women's Christian Temperance Union (U.S.) A movement for Puritanism and prohibition at the beginning of the 20th C. They were for prohibition of alcohol and tobacco as an improvement to society's morals. American Heritage (1961)

15.

MODES OF TOBACCO USE

"When pipe in mouth I may hide myself behind in smoke,
Oh brother—that is bliss!
I would not change places with a Knight then.
I see the whirling smoke, and it amuses me.
Yet—are not actually all things here on earth like smoke?"
(from a student at Grouigen University, 1816)

PIPES

The Raleigh Pipe and the Gouda Pipe, 17th century

Pipe smoking had been one of the modes of tobacco use by pre-Columbian aborigines. The Indians of North America smoked pipes, and that was the reason why the British and Dutch took up pipe smoking rather than cigars, as did their Spanish counterparts. The Thirty Years War in Europe facilitated the practice and spread of pipe smoking through Europe. By the 17th century pipes had become the predominant method of smoking around the world thanks to the influence of England and Holland. Most 17th-century pipes were made of clay and cost about a penny a pipe, but other materials such as silver, porcelain, meerschaum, briar, and corncob were used and eventually displaced clay pipes in the 19th century. As time went on, pipes became more elaborate. They were made in a variety of sizes, with long-stemmed pipes more popular in America and Asia than in Europe.

Why pipe smoking did not catch on in Spain is understandable because in Hispaniola, Cuba, and in parts of México tobacco was smoked mainly in cigars. These cultures also smoked tobacco in pipes or chewed it, but it was smoked mainly in the form of rolled leaves. The Spanish colonizers of the New World brought the tobacco habit back to Spain mainly in the form of cigars. The Spaniards had the largest cigar factory in the 17th century, located in Seville.

1. Age of the Pipe
Smoking and "the age of the pipe," 16th-17th centuries, and how tobacco was used in different parts of the world.

England & Holland	16th-17th centuries	The Raleigh and the Gouda Pipe (Clay)
Africa	17th-18th centuries	Wood, Clay and Ivory Iron Pipes
Germany	17th-18th centuries	The Meerschaum Pipe
England	17th-18th centuries	The Briar Pipe
India	18th-19th centuries	The Hookah "Hubble-Bubble" Water Pipe The Egyptian, Narghile Pipe
Arabia	18th-19th centuries	Shisha (Of Beautiful Elegance)
Turkey	18th-19th centuries	Narghile and Chibouk Water Pipes
Persia	18th-19th centuries	Narghile and Chibouk Water Pipes
Lottman Afghanistan Tchilim	18th-19th centuries	Narghile and Chibouk Water Pipes
Japan & China		Pipes with Small Bowls, Kisseros

Supplements to the pipe, snuff, cigars, and cigarettes:

Pipe Cleaners & Holders
Snuff Boxes & Handkerchiefs
Cigar Holders
Tinder Boxes & Flint "Guns"
Matches
Special Chairs and Divans

2. Age of Rappee (Snuff)	18th century
3. Napoleonic Era Age of the Cigar	19th century
4. U.S. Age of Cigar and Chewing	19th century
5. Age of Cigarette & Cigar	19th & 20th centuries

Little tube of mighty power
Charmer of an Idle Hour
Object of my warm desire
Lip of wax, an Eye of Fire . . .
Can afford his tube to feed
With the fragrant Indian Weed
Pleasure for a nose divine
Incense of the God of Wine."
Ambrose Phillips

Europeans started smoking the Raleigh Pipe at the end of the 16th century. Because of the enmity between King James I and Sir Walter

Raleigh, however, the Raleigh pipe became a symbol of defiance and a thorn in the side of King James I, who disliked the "filthy weed" and persecuted pipe smokers.

Pipe Smoking and Manufacturing of Pipes in England and Holland

The long, clay English pipes were called "church workers" before they were called "aldermen," a title that gave them high social status. The poor, who could not afford to buy large amounts of tobacco, smoked miniature "elfin" pipes in Scotland or "fairy-pipes" in Ireland. The English smoked the Raleigh Pipe. (Fig. 60) During the American Revolution, exports of tobacco to England from her rebellious American colonies dwindled, making tobacco scarce and very expensive, almost prohibitively so for the poor. The rich had no problem in getting tobacco at any

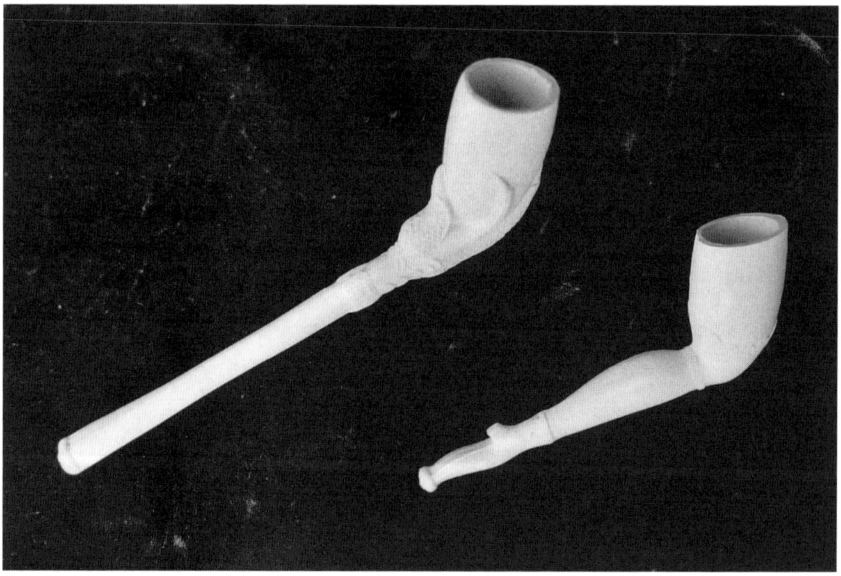

Fig. 60 Raleigh Pipes. Cheap and disposable, these pipes and the Gouda Dutch pipes were the most popular pipes in the 17th C. Private collection of R. Oropeza.

price. People then as now managed to find a way to keep on smoking. Tobacco had become a priority in their daily lives. Tobacco barons eventually controlled the business. Mansions were built in George Square, England, and the streets were named after places where tobacco was grown, such as Virginia and Savannah in the United States.[95]

Historical events made Holland an important center for pipe making. English soldiers had begun migrating to Holland after the death of Queen

Elizabeth in 1603. They were persecuted by King James I because of their Puritan religious beliefs, and also because he was against smoking.[96] Beginning in 1609, the soldiers of Prince Maurice started a pipe making business in various towns in the Netherlands. The English were up to this time the best pipe makers, but now Holland became the leader in pipe making. The Dutch Gouda pipe (1647) became the most popular pipe in Europe. Tobacco shops were created for the sale of tobacco and coffee. These *tabagies* still remain in Holland and Denmark, where now they sell hashish and other stimulants as well as tobacco. The Dutch had their own tobacco plantations[97] and factories where they improved the manufacturing of tobacco and tobacco products. The Dutch have always been proud of the high quality of their products.

Smoking had become a major problem in Holland by the end of the 17th century. Everyone smoked a pipe. An attempt was made by the authorities to change the smoking habits of the Dutch. Boswell censured smoking in 1773: "Don't smoke any more because it makes you sick and a foreigner need not do it." After living only a short time in Holland, however, Boswell was of a different opinion. He had become addicted to tobacco and composed the following verse: "With the same ease the black gourds feed on tripe have I, James Boswell, learned to smoke a pipe for I am now a very Dutchman grown."

By 1770 chewing tobacco and inhaling snuff had become fads. Pipes were smoked less frequently and the Gouda factory, as well as other pipe factories, went broke.

Water Pipes

"His gurgle was the last thing that I
heard when I fell sleep, and the first
when I awoke in the morning."[98]

Fenier—241

The Narghile, the Hookah ("Hubble-Bubble"), the Manipur, the Dafta, the Tchilim, and the Chibouk

Water pipes had their origin in Africa, where they were used not only for smoking hemp (Dakka), but also tobacco. Dakka had been smoked in water pipes long before the introduction of tobacco to Africa in the 16th century. The Narghile pipe originally was made from a coconut shell half-filled with water, with two pipes or reeds and a stem (*narghile* in arabic means "coconut"). The smoke had to pass through the water to cool the smoke in the tropical climates, but water makes tobacco lose its scent so it had to be flavored with scented herbs.

Ottoman Empire

A crude water pipe was introduced to Arabia from Africa by slave traders long before the introduction of tobacco. In Arabia they smoked tobacco in water pipes that they called Narghile or Shisha. In India they called it a *hookah*. Some of these pipes were beautifully elegant. In Afghanistan they smoked water pipes that were called Tchilim. This pipe became very popular in Africa (especially Algiers) where it was smoked not only by men, but also by women as we can see in this painting at the Louvre by Eugène Delacroix. (1872) Muslim men and women were prohibited from smoking by law, but they smoked anyway.

The Turks had special rooms and special seats (divans) to smoke their Narghile and Chibouks (long pipes) filled with a mixture of aloe, rose leaves, opium or hashish (and God knows what else) mixed with tobacco. (See Fig. 63)

"Sublime Tobacco! Which from East to West
Cheers the tar's labour or the Torchman's rest;
Which on the Moslem's Ottoman divides
His hours, and rival opium and his brides;
Magnificent in Stamboul, but less grand,
Though not less loved, on Wapping or the Strand;
Divine in Hookas, glorious in a pipe.
When tipped with amber, mellow, rich, and ripe;
Like other charmers, wooing the caress,
More dazzlingly when daring in full dress;
Yet thy true lovers more admire by far
Thy naked beautifies—Give me a cigar!"
Lord Byron, *The Island*

Indian Water Pipes

The hookah, or "Hubble-Bubble" water pipe for those who smoked it, was the necessary indispensable tool dispensing cool gurgling pleasure everyday. "At the Bles hour when nicotine and coffee, the day's first cup and smoke unite and soar into the brain, with sleep still fluffy, like Bee and Queen on their nuptial flight."[99] This pipe was smoked with gusto and was frequently used as a sign of wealth during the Mogul Empire, (India) as well as by the elite Indian classes during British rule.

Some beautiful ornate pieces remain in displays at the Victoria and Albert Museum in London. Some were made of inlaid metal or colored glass as depicted in Fig. C34. This type of pipe was ideal during social gatherings to initiate camaraderie or just simply to kill time. The flexible long cord with the ivory mouthpiece was passed around. Smokers inhaled the tobacco without touching their lips to the mouthpiece, by cupping their hands. Small, portable hookahs for males and females were designed for long trips either on horse or foot.

Water Pipes in Egypt

In Egypt tobacco was smoked in cigars or pipes, but Narghile water pipes were preferred. One still sees this use of water pipes in the towns close to the banks of the Nile, (Fig. 61) as well as in every other Arab country,

Fig. 61. Smoking a Narghile on the banks of the Nile (1999). Photography by R. Oropeza. 1999

Fig. 62. Advertising sign for Narghiles in a tobacco shop in Granada, Spain. Photography by R. Oropeza. (2004)

in Granada, (Fig. 62) Spain, and even in Queens, New York, where Moroccan and other immigrants from the Middle East congregate.
Scented tobacco is sold in these shops. (Fig. 63)

Fig.63. Box of Scented Tobacco. Granada, Spain. Photography by R. Oropeza. (2004)

A Story of the Pasha of Egypt and His Long ("Chibouks") Pipe

An Englishman, Robert Carzon, mentioned that while he was on a business trip to Egypt he was received ceremonially by Pasha Mehemet Ali (1805–1849). After receiving the elaborate oriental compliments, "they were given cups entirely covered with large diamonds. A pipe was then brought to the Pasha, but not to us. This pipe was about seven feet long, the mouth piece of light green amber was a foot long, and a foot more below the mouth piece, as well as another part of the pipe lower down, was richly set with diamonds of great value, with a diamond tassel hanging from it. . . . When the Pasha was young he had been a modest tobacco trader but with time he had increased his wealth so he must have had a feeling of importance."

Narghile Water Pipe

Thackeray described his experience smoking from a Narghile pipe while on a cruise ship in passage to Istanbul in 1846: "You are laid gently on the reposing bed, somebody brings a Narghile which tastes as tobacco must taste in Mohamed's paradise. A cool sweet dreamy languor takes possession of the purified frame; and half an hour of such delicious laziness is spent over the pipe as is known in Europe."[100]

Chilim (Tschilim) Pipe

"Afghans on the road are always stopping to gossip, to drink or to pray ... they seldom travel without their chilim for it is a punishment to them to pass a single hour without smoking."[101]

Chinese Bamboo and Metal Water Pipes and Japanese Kisseros

The Chinese mixed tobacco and opium in their ornate metal water pipes (Fig. 64) or in regular, long-stemmed bamboo pipes with a small bowl. (Fig. C40) Tobacco was popular in China in the 17th and 18th centuries in spite of decrees against smoking and punishments such as being paraded in front of a crowd and forced to shout that nobody should do as they had done. Some repeat offenders were severely punished, even executed, as in Turkey during the reign of Murad IV. The result was the opposite, however; more Chinese started to smoke. At the end of the 19th century, the Chinese were changing over to cigarette smoking like everybody else in the world.

Revival of the Pipe in the 19th Century, The individual Pipe

The Meerschaum Pipe
Germany, Austria, Turkey

Smoking clay pipes with a short stem in the 19th century was for the working class. The upper class smoked pipes with longer stems. The pride of the higher class became the Meerschaum pipe, just as the water pipe or hookah had been the pipe of the nobles in India and the Middle East. Meerschaum is a foamy stone of variable quality that is found floating in the Black Sea. These pipes are still masterfully crafted in Turkey. A pipe can be carved out in two hours on the spot while you wait. (Fig. C35)The bowl of a Meerschaum pipe absorbs the juice of the tobacco, giving it a special peach hue that made it more desirable and increased its monetary value. These pipes made their appearance in Europe at the end of the 18th century and became very popular in the 19th century. (Fig. 65)

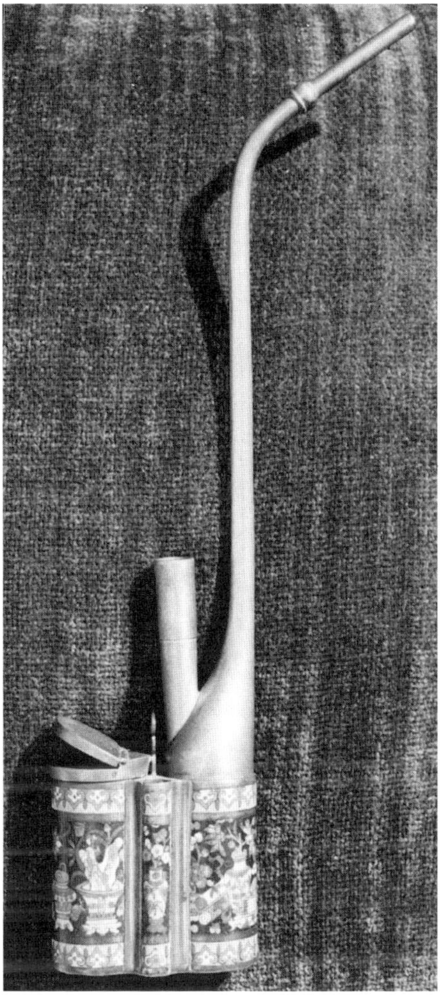

Fig. 64. Chinese cloisonné painted and emaneled bubble pipe used for smoking both tobacco and opium (vertical cylinder). Private collection of R. Oropeza.

The Briar Pipe—Brugère "Health" **England**

This pipe was made from the tough roots of the southern bruyere. (Brugère-Erica Arborea) The pipe was simple, strong, and beautiful when well polished—very English indeed. Most briar pipes were not decorated, and

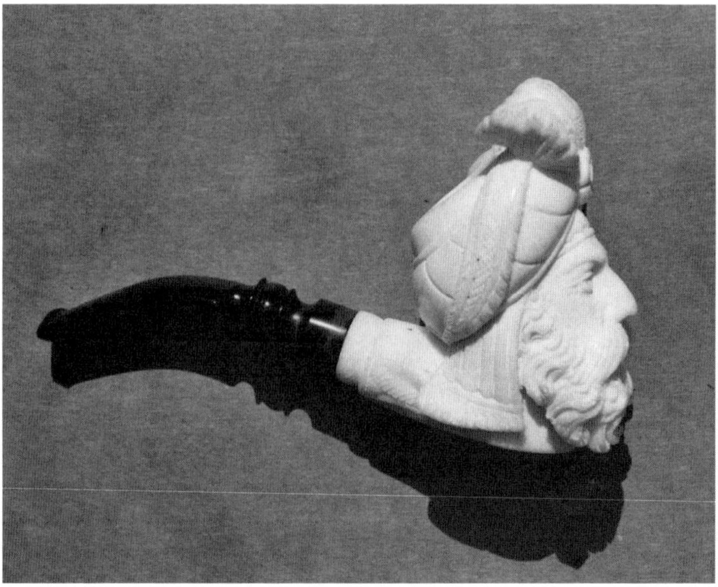

Fig. 65. Finished carving of a Meerschaum Pipe (2 hrs). Capadoccia Turkey (1999) Collection R. Oropeza.

only used for special occasions. They were made mainly in Salisbury, England. They are kept in museums such as the Winchester Museum and the Dorchester Museum in England. The Briar Pipe became the most popular pipe in the world for the next century, and was the inspiration of painters.

Albert Einstein puffed with gusto and inspiration, on his Briar Pipe, while developing his theory $E=mc^2$ and whiffed silently and pensively while listening to Leo Szilar to explain his chain reaction theory and the need to write a letter to President Roosevelt requesting funds to develop the A bomb.

After Szsilar had completed his long explanation about his theory, Einstein stopped puffing on his pipe and murmured ''why I didn't think of that.''

After Szilar had convinced Einstein to write that letter to Roosevelt, he went back to New York. In the meantime Einstein had gone sailing on his small boat on the Peconic Bay, carrying along his pipe, his always faithful companion. (from Peter Bucky son of Gustav Bucky, Albert Einstein's best friend.)

All this happened in Cutchogue L.I., NY during his summer vaction.

It was the favorite also of Douglas MacArthur while chasing Pancho Villa in the mountains of México in 1916. When MacArthur was evacuating Leyte in the Philippines during the Second World War, he clenched on his corncob pipe and muttered, "I shall return." This pipe was his pacifier while under stress. When the Korean War broke out in 1950, a reporter asked MacArthur to assess the state of the conflict while he was going into the scene of the action. "With a mist of anxious thought and discussion, he quickly got up to rummage in his bag for a cherished corn cob pipe as though turning to an oracle!"

An entire industry was created for the tobacco pipe in the 19th century. Pipes of all kinds and quality were made, and with them tobacco boxes to keep the tobacco fresh. These boxes varied from simple to highly ornate, and were made of gold, silver, turtle shell, tin, or wood. Some of them depicted natural scenes or historical moments, or were engraved to capture a special family occasion. Often they were custom made.

Tobacco pouches were made to carry tobacco. Some of them were simple, but some were richly embroidered. Cleaners for the bowls and stems of the pipes varied from the simple toothpick-like cleaner to more expensive cleaners made of ivory, silver, or gold. There were carved pipe holders of wood, ivory, ceramic, leather, and so on. Standpipes of different types were made to hold special pipes with long stems. Lights for the pipes were developed so that smokers could light their pipes without the inconvenience of bending over to reach for a live coal. The tinderbox was invented. The English flint was made of steel and wood in the form of a pistol. Some were engraved. Special smoking chairs were made. These chairs had special places to keep not only the pipes, but also all the needed paraphernalia.

THE AGE OF SNUFF

"Bend with great pomp
Two feet backward in space
Gracefully forward again
Now bend without disgrace
According to the manners of the French
Now take from the pocket of our camisole
As though you did it at the command
Thought you at the French School
Your snuff box, small or middling,
Tap it well mannered,
Snuff with a noble grace,
Open it quickly
Offer it to your friend."
An old poem

Fig. 66. Demoiselles by Boilly taking a dainty pinch of snuff. (18th c.)

"Every man, woman and child took snuff inevitable."[102] (Fig. 66)

In Europe the fad of taking snuff started during the last two decades of the 17th century while pipe smoking was still going strong. This was the most disgusting way to use tobacco, but at that time it was well accepted at all levels of society. The Scots even plugged their noses with tobacco! Snuff was viewed as more elegant than a pipe by the end of the 18th century because the smoke from pipes offended the upper classes. Another factor that led to the popularity of snuff was that people who did not have a strong palate could not smoke the unblended tobacco available at that time. It was simply too strong for many. In the first half of the 19th century, snuff became the most desirable mode of tobacco use for both men and women. The aristocracy of France, Spain, and England, then most of Europe, left pipe smoking to the lower classes and engaged almost exclusively in the practice of snuff-taking.

The clergy used snuff as far back as the 17th century because snuff could be easily concealed and was inconspicuous, even during the ceremony of the mass. Pope Urban VII was a heavy snuff user. Once he offered snuff to one of his clerics, who declined it saying, "Your Holiness, I don't have that vice," to which the Pope replied, "Were it a vice, you would have it!"[103]

It was fashionable and debonair at parties given by the French aristocracy at the court of Catherine de Medici, to carry fancy snuff boxes and engage in this practice. Sneezing was "chic" and part of the culture. A sneeze would break the ice and provide a lady with the chance to show the bracelets on her wrist and the rings on her fingers while she cleaned her nose of the peach-hued fluid.[104]

Snuff never took as strong a hold in the United States as in Europe, but it was popular all the same. Small snuff mills were started in Virginia in 1730 and in New York in 1760 by the Lorillard Company, which is still in existence. The American Snuff Company, W.E. Garret and Sons of Memphis, Tennessee (1782) produced 4,917,418 pounds of Scotch Snuff in 1900, and 40,645,689 pounds in 1930. Snuff became the first tobacco product to be branded in the United States.

The chief manufacturers of snuff were in England, Scotland, France, Holland, and Germany. However, the Spanish snuff from Cuba, known as "musty," was considered *la crème de la crème* for its scent and humidity. Snuff manufacture included an elaborate process of washing and grinding, and the addition of colors and scents to please all customers. Snuff induced sneezing, so the fancy handkerchief had to be invented. Because the handkerchief had to be handy at all times, special snuff shirts were created with pockets to hold the handkerchiefs.

Manufacturers made individual custom snuff boxes out of different materials. Elegant snuffboxes[105] were the preferred gift of the 18th century, especially among the aristocracy and royalty, who often put jewelry or money inside the box. Scottish snuff boxes became some of the most popular. Marie Antoinette received fifty-two snuffboxes as wedding presents. For the common people, the ordinary wood box would do.[106] For the aristocracy there were certain rituals that had to be followed when taking snuff.[107]

The Irish in the 17th century called snuff "smutching," and they used snuff to "purge their brains"! In France snuff was called "rappe," derived from the word *rasp*, the tool used to grate the tobacco leaf.[108] Rappe is also a brand name for a French-made snuff.[109]

Queen Anne brought the French fad of taking snuff to the English throne, where it was soon accepted by the English court and the aristocracy. George IV spent £8,205 on snuff boxes for gifts to his foreign ministers on his coronation.[110] Princess Margaret of Austria gave a beautiful snuffbox to Franz Joseph Haydn after he dedicated a symphony to her in the 18th century. Later, the great English statesman Winston Churchill was fond of taking snuff before he became a cigar smoker.

Madame de Pompadour "invented" her own brand of snuff, called "Essence du Tabac de Pompadour,' which claimed to improve memory.

Napoleon Bonaparte consumed seven pounds of snuff a month. When exhausted or stressed, he took a copious pinch of snuff. In 1796, during the Italian campaign, Napoleon used a snuffbox with the portrait of his beloved Josephine. When the lid of the snuff box got broken, he considered it a bad omen (he was very superstitious) and asked Josephine to send him another box with her portrait. Napoleon also carried another snuffbox with the portraits of Regulus, Sulla, Pompey and Caesar.

Napoleon disliked pipe smoking, but like other statesmen found it convenient to accept tobacco use in his country because of the revenues from the taxation of tobacco. He looked the other way when his generals, like Vandamme, Lasalle, and especially Field Marshall Prince Blücker Von Wahistatt smoked their pipes.[111]

While he was campaigning in Egypt in 1796, Napoleon was given a Narghile pipe as a gift from a Persian envoy. When he inhaled it deeply and smoke had reached his lungs he had a violent coughing spell and he is said to have exclaimed, 'Take it away! How foul! You pigs! It makes me sick.' To relax, he then turned to his snuff box and took a good pinch of it.[112]

There is another legend that one day when Napoleon was presiding over his council in his usual strong, imperial manner, he kept opening a small, gold snuff box to have a pinch of snuff, which he took elegantly and almost inconspicuously. Some of it spilled on his lapel, and at the end of the meeting his lapel was full of snuff. He handed his empty snuff box to his page, asked him to refill it, and then quietly shook the snuff from his lapel.

Napoleon carried a box of snuff and snorted at the battle of Waterloo. Meanwhile, the Dutch and Flemish troops in Wellington's army wanted to be inconspicuous and went behind trees to sneak a smoke.

Clayden tells us that on Napoleon's return from imprisonment on the Island of Elba, the Emperor ate "a little dinner . . . and after, a beaucoup de tabac." While imprisoned by the English the second time, on St. Helena, a rocky, Atlantic Island of England, his addiction to snuff increased. He took generous pinches of his favorite snuff from his tortoise snuff box during his daily walks with his appointed companion generals and his physician, Barry E. O'Meara, in spite of his constant, upper abdominal pains. He died of cancer of the stomach.

O'Meara, Napoleon's last physician on St. Helena, wrote one of the last first-hand biographies of Napoleon in 1822, the last one was by Count de las Casels, 'Memorial de Saint Hélène. Napoleon had become very attached to this physician in much the same way that he became attached to Dr. J. Larray, the first physician to bring emergency services (ambulances) close to the battlefront while Napoleon was campaigning in Russia. Napoleon lived at Longwood House in St. Helena, where a detailed record was kept of his daily activities. Before Dr. O'Meara departed from St. Helena to France, Napoleon instructed him as to whom he should contact—his wife Louise, his son, and others—then he presented O'Meara with a superb snuff box and a statue of himself and said, "Good-bye, O'Meara, we will never see each other again. Good luck." Then he took a large pinch of snuff and walked away.

There were those that opposed the use of tobacco in any form, including snuff. They found snuff repulsive and in poor taste. Franciscus Phillippus Florinus wrote in 1750: " . . . but now things have come to such a pass (Lord have mercy on us) that the poor people of the old world get something from the new world which might soon pack us off into the next world. Especially is this true of the beloved herb tobacco, a noble plant, but so missed among us."

The Electress Sophia of Hanover said: "This snuff is a disgusting affair. I hope you do not take any, Dear Louis! It puts me in a temper when I see all the women here, with dirty noses as if they had rubbed them in the muck—excuse the word!—come and stick their fingers into any man's snuff box. It makes me sick, it is so disgusting."[113]

The Pacific Islanders were viewed with dislike by Mark Twain during one of his trips. According to him, "They smoked, chewed and sniffed snuff, conversed and spit on each other." Twain found it disgusting, but he was never without his own pipe or cigar.

On the other hand there were those that attested to the value of snuff, even claiming that it had miraculous properties. One gentleman said, "Sir, I was born stone blind and so continued up to my seventieth year when on taking one small pinch of your infallible, miraculous, regenerating, penetrative, sight perpetuator, my eyes opened strong and clear . . . more brilliant than the optic of Hebe."

While living in Augsburg, Germany, in 1777, W.A. Mozart wrote a letter to his father complaining that he had been insulted by a Baron Bagge and an army officer because he had received a decoration from the Pope. The baron and the officer were passing pinches of snuff to each other, and they passed a pinch to Mozart who said, "It would be easier for me to obtain all the orders which it is possible for you to win than for you to become what I am, even if you were to die twice and be born again. Take a pinch of snuff on that!" Finishing his remarks, Mozart put his hat on and walked away.[114] (The Letters of Mozart by E. Anderson 1985)

A story goes that a traveler, Lawrence Sterne, was feeling guilty for being impolite during his travels in France to a monk named Father Lorenzo. "The monk met him again and offered him a pinch from his fancy horn snuff-box. (Fig. 66) Sterne in return presented him with his

Fig. 67. Fancy English horn-pewter snuff box 18th century. Private collection R. Oropeza

own superior tortoise shell box. They had a friendly chat, and at the end they exchanged their boxes. It remained always in his hands until the end, until he sat on Father Lorenzo's grave."[115]

There was snuff made especially for the morning, for the afternoon, and for after dinner. There was snuff for men, for women and, I suppose, also snuff especially made for fools.

Snuff was very popular in most of Africa. Missionaries often carried snuff while on their mission "to save souls" through conversion or coercion . . . or by sharing snuff with the natives. These missionaries consumed large amounts of snuff, sniffing deeply while rolling their eyes, even at the time of conversion as if trying to put a spell on the natives!

A Zulu epic chanted while going to war in the 19th century mentions snuff in a symbolic way as if snuff had a life of its own. The Africans of the south eagerly took to tobacco in the form of pipe smoking and snuff taking. They often smoked tobacco either in long pipes made of reeds with earthen bowls, or with fanciful iron stems shielded with ivory. They had a small calabash (gourd containing snuff) hanging from their necks and a special spoon ready to be used. One account of the Zulus mentions seeing them "wearing wooden tweezers on their noses. On inquire we discussed that they injected snuff mixed with water, and then put the apparatus on to keep the concoction from wasting away at once."[116]

Chewing Tobacco
United States, 19th century

" . . . made Americans to chew
on their cigars as well as
sucking them . . . like a
burning candle in both ends."
Trillope's Mounse

Chewing tobacco was not a new practice started by the Europeans. The Indians of America chewed or sucked tobacco long before Columbus. It was a common practice among the shamans during their healing and religious ceremonies, and it is still a common practice among the tribes of the Orinoco and Amazon Basin and the southeastern United States. Europeans and Americans took to this practice at the end of the 19th century.[117] Chewing tobacco became very popular in the United States during the American Civil War. It was a part of the army rations and "the good old twist" was encouraged by commanders. It was convenient.

There was no need for fire to light it, and it was immediately available in moments of stress or during battle. The habit disgusted visiting Europeans who were appalled at the Americans' practice of spitting tobacco juice everywhere, which was an acceptable practice then. Of necessity the spittoon was invented. Nineteenth-century Europeans grudgingly admired the accuracy with which Americans could hit spittoons—or any other target they chose, and that led to spitting contests.

Chewing tobacco became more popular in the United States than in Europe. Americans preferred chewing over sniffing, which explains why most tobacco growers in the U.S. concentrated at that time on the production of chewing tobacco. To make chewing tobacco palatable, sweeteners like sugar or licorice were added.[118] Chewing tobacco was packed in plugs and twists. Paraphernalia associated with tobacco chewing were tobacco pouches and spittoons. Today Americans are still very fond of chewing, but they mostly chew gum rather than tobacco. Some baseball players still chew tobacco, and the practice is still popular among sailors, outdoor laborers, and cowboys but it is not well accepted these days to chew tobacco in public.

Even before the 20th century, the habit of chewing tobacco was seen by critics as having some serious (and disgusting) drawbacks including bad breath, blackening of the teeth, a diminished sense of taste, constant salivation, drooling, and the incessant spitting in public. Charles Dickens reported with disgust that when he came to America he had the misfortune of traveling close to American tobacco chewers, and he was depressed to see the carpet of the United States Capitol squirted with spit. According to Dickens, Washington, D.C., should have been called the headquarters of tobacco-tinctured saliva.

Spitting Contests

"There is a story that in the silver mines of California, in the Ghost Town of Calico, a notice still hangs, recording a winning effort on the spitting competition. There were spitting competitions as a pastime."[119] In the same town, "the spittoons that were present in town were not always used. People that were disgusted with this practice put a sign up asking chewers to spit into one another rather than on the floor."[120]

There is a more recent story of a 1946 spitting contest in North Carolina. "The most energetic 'juice flinger' in the 1946 tournament was unable to match by almost six feet the long distance squirt of twenty-one feet, two inches obtained during the 1945 competition."[121]

Donkin tells us that in the town of Regina in Canada there is a big wooden hotel where "long-faced men sit silent around the stove, only varying the grim monotony by an occasional expectoration of tobacco juice. Sometimes they may break out and engage in the congenial pastime of swapping lies."

Chewing on a cigar was also very common in America. One European observer remarked about cigar smoking: "It was like burning a candle on both ends. The cigar burned very quickly on one end and was chewed fast on the other end."

Age of the Cigar
19th-21st century

"Sublime Tobacco! Which from East to West . . .
Like other charms, wooing the caress
More dazzling when daring in full dress;
Yet they true lovers more admired by far
Thy naked beauties—give me a cigar."
Lord Byron, *The Island*

Horace Greeley defined the cigar "as a fire on one end and a fool on the other." In the 20th century the cigar became a familiar companion of many entertainers on radio, in movies, and on television—Groucho Marx, Milton Berle, Joey Bishop, Jimmy Durante, Robert Duval, Bill Cosby, James Coburn, Robert DeNiro, Charles Chaplin, Alfred Hitchcock, Jean Paul Belmondo, Edward G. Robinson, Paul Newman, Arnold Schwarzenegger, Humphrey Bogart, and many others.

From the time that Spain began to colonize the New World in the 16th century, the Spanish colonizers became cigar smokers and brought the habit back to Spain. Spain had the largest cigar factory, in Seville, where practically all workers were females (who also smoked). This was the inspiration for *Carmen* by Bizet. Most women elsewhere did not become ardent cigar smokers, preferring the habit of snuff taking instead. Spaniards remained faithful friends of the cigar even at the time when England, Holland and the rest of Europe preferred pipe smoking. Richard Ford said that a "Spaniard without a cigar in his mouth is like a house without a chimney or a boat without a funnel."

An observer in Jamaica in the 19th century mentioned a trial of Spanish pirates. They were all splendid looking fellows, according to his report, "every Spaniard has a beautiful mouth until he spoils it with the beastly cigar, as far as his well-formed firm lips can be spoiled, but his teeth he generally does destroy early in life." In México the stereotypical Spaniard is depicted as wearing a black béret and clenching a cigar in his mouth even while speaking. This is a holdover from México's fight for independence from Spain in 1810–1820.

During the Napoleonic occupation of Spain in 1808, the French took up cigar smoking and helped spread this custom to other parts of Europe. The Peninsular War with England in Spain in 1815 had the same effect. England imported 26 pounds of cigars in 1826, but by 1830 the figure had jumped to 250,000 pounds.

The first cigar factory following the model of the one in Seville was built in Hamburg in 1788 by H. H. Schottmann. By 1793 it was common to see Germans smoking cigars, which they called *zigarro*. In Austria they were called *zigari*. Paul Leforge tells us that "it was not commonly sold in the streets, but an expensive cigar was a hallmark of the aristocracy. Dirty old pipes were considered the hallmark of riff-raft."

When this habit became rampant in Germany, an attempt was made to stop its use. During an epidemic of cholera in the 18th century, however, there was no effective medication against the disease because its cause was not understood, and tobacco use was not only allowed, but even encouraged by the minister of the interior (health minister), Baron Von Brenn.

Louis Napoleon III of France, an ardent smoker of cigars and cigarettes,[122] waged a war of occupation in México in the middle of the 19th century in an attempt to establish a monarchy for Maximilian and Charlotte. During the short decade of occupation in the 1860s, the French troops were exposed to the practice of cigar smoking that was prevalent with their Mexican counterparts. (Fig. 68) This gave another boost to cigar smoking in France.

Smoking cigars was a sign of distinction and prestige in aristocratic México City and Paris at the end of the 19th c. (Fig. C.36)

United States cigar smoking was greatly influenced by the exposure of soldiers to the habit during the Spanish-American War in Cuba at the end of the 19th century. Before then some soldiers had taken up smoking during the American occupation of México in 1846–1848.

General Ulysses S. Grant was an ardent cigar smoker. He was named "the father of cigar smoking" during the Civil War because he was always seen smoking a cigar at the battlefront (he was even seen with a cigar in his mouth while sleeping).[123] After the capture of Fort Donelson in 1862, the states of the Union presented Grant with a gift of 11,000

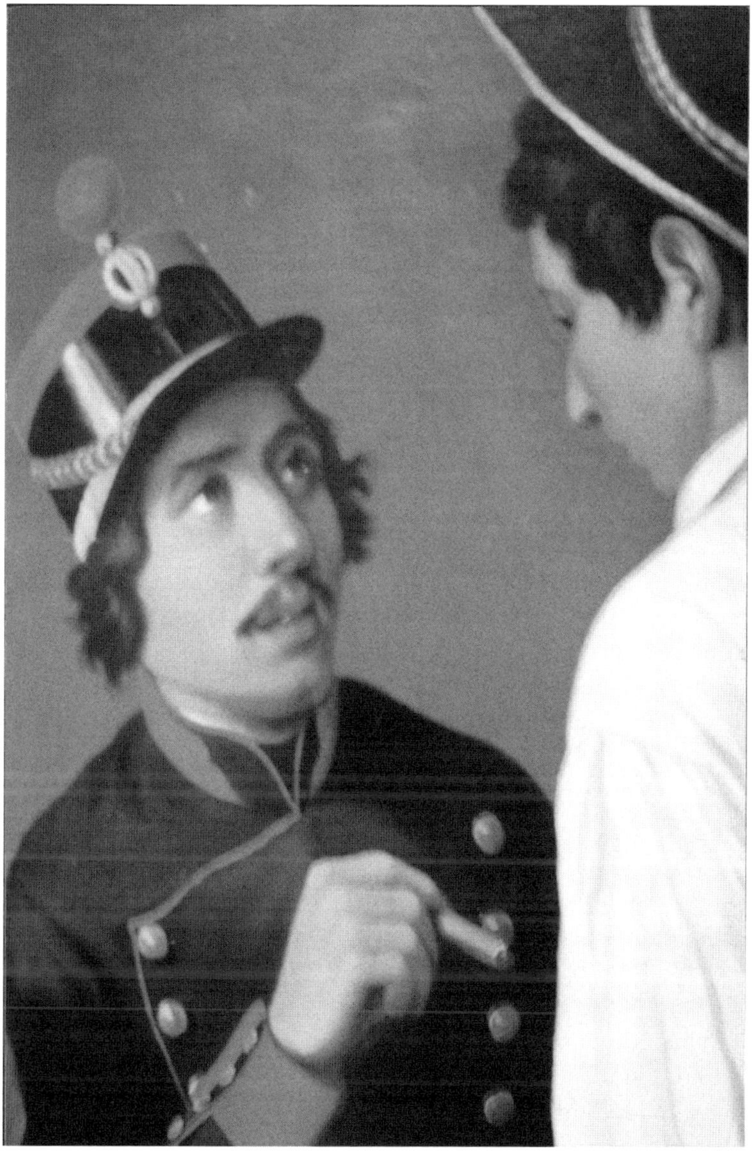

Fig. 68. A soldier of Napoleon III's army sharing a cigar with his Mexican counterpart in 1861. From an exhibit of 19th Century México City, Photography by R. Oropeza.

cigars (which might have been the best revenge of Confederate General Lee, who was not a smoker). When Grant found that he could not give away all these cigars, he decided that he would try to smoke them all instead. Grant believed, as did others at the time, that cigar smoking sharpened his wits and helped him to win the Civil War. Thomas Edison also thought that smoking enlightened his brain. Paradoxically, however, he was against his workers smoking in his own factory.[124]

Grant sometimes also smoked a pipe or a cigarette to rest his mouth after clenching on his cigar the whole day, but he was never without tobacco. There was nothing that would please him more than a gift of good Cuban cigars, and he had tons of them in his office, his bedroom, his desk, and every place else. Grant eventually paid a heavy price for this indulgence. He developed cancer of the larynx, (throat) practically choking to death during the last days of his life. He died of this cancer soon after he had finished his memoirs.

American presidents are supposed to be role models for the average United States citizen, but this of course has not been the case. Most presidents of the United States have been tobacco users, and some of the founding fathers were tobacco planters in Virginia, such as George Washington and Thomas Jefferson. President John Adams and his son, J. Quincy Adams, James Madison (whose wife was a snuffer), Andrew Jackson and his wife—all smoked at the White House, as did Ulysses Grant, Chester Arthur, Benjamin Harrison, William McKinley, William Howard Taft, Calvin Coolidge, Franklin Roosevelt, Dwight Eisenhower,[125] and John Kennedy. Kennedy reportedly said to his cigar-smoking press secretary, Pierre Salinger, during the invasion of the Bay of Pigs in Cuba, "I need a lot of cigars." Salinger asked, "How many, Mr. President?" Kennedy answered, "About a thousand."[126]

Richard Nixon also enjoyed cigars and was the last president to offer them to his guests at the White House.[127] Carter did not smoke or drink. Gerald Ford smoked a pipe and Reagan chewed tobacco, but Bush (Sr.) did not smoke. Clinton smoked cigars on the golf course because he was not allowed to smoke them in the White House. President G. W. Bush does not smoke.

Many other politicians in the United States and abroad have smoked cigars, cigarettes or pipes. Tobacco crosses political boundaries and is enjoyed by people of different political ideologies dictators like Fidel Castro or revolutionaires like Emiliano Zapata. (Fig. C37) The Russian dictator Nikolai Lenin was a cigar smoker. It is said of Lenin that he was in such a state of poverty in Switzerland that he stole cigars from the tobacco shop of Davidoff because he could not pay for them.[128]

In Imperial England during the 19th century, Sir Edward Benthall was having lunch with a jute magnet, the Minister of Labor of Bengal (India), H.S. Suhrawardy. After enjoying the numerous courses of the meal, magnificent cigars were offered to the dignitaries. This reminds one of the "after dinner" smoke of Motecuhzoma on his encounter with Cortés in 1519, when Motecuhzoma offered his guests gilded pipes at the end of the banquet.

Winston Churchill had cigars specially made for him. The cigar became a symbol of Churchill's defiance during the bombing of London by the Germans in World War II. Photos were taken of him making the V sign with his fingers while clenching a cigar in his mouth and walking defiantly through the rubble of the city after a night of bombing by the Germans.

Emperor Bismarck of Germany was a heavy cigar smoker. As a sign of defiance against Count Thun of Austria, an adversary, Bismarck smoked a pipe when he was not supposed to. He was seen constantly clenching his cigar during the war with France in 1866–1871. "During the signing of the peace terms he offered a cigar to his counterpart General Faure, which he declined. Bismarck then commented, 'You are wrong. When a man begins a discussion which may easily lead to a heated argument or even a show of temper, it is always better to smoke while one is talking. You see?' and he lit his expensive Havana cigar."[129] The cigar in the 20th century became a symbol of elegance, success, power, and prestige, a symbol of status for the famous and the infamous. Movie actors and comedians, musicians, physicians, presidents, politicians, and kingpins in the mafia underworld shared the weed. Even a pauper felt momentarily important with a cigar in his mouth.

By the end of the 19th century, America was importing many of its cigars from Cuba, which had—and still has—the best cigar tobacco. Cigars have been handmade for more than two hundred years by master craftsmen in Havana. They are workers of such a distinction that they have been called "intellectuals of the proletariat." The wood used for the cigar boxes is made of cedar, which repels all insects. Famous Cuban brands are immediately recognized by their names, such as Larangos, Garcías, Partagas, and Cohibas. Some Cuban cigar makers emigrated to Miami, to the Dominican Republic, or to Central America after the Castro revolution in 1959 and started tobacco manufacturing companies in their countries of exile. English and Dutch manufacturers have always been proud of the excellent quality of their own cigars, but they were never as good as the Cuban "Cohiba" smoked by Fidel Castro.

Sigmund Freud experimented with cocaine while his patients lay on the couch and he developed his concept of psychoanalysis. To have a break between patients, he puffed on his Habana cigar.

The cigar had triumphed over the antitobacconists.

" . . . some want a German row
some wish a Russian War.
I care not—I am at peace,
So I have my cigar."
Thomas Hood, 19th century

16.

THE INFLUENCE OF TOBACCO IN THE LITERARY, ARTISTIC, AND MUSICAL WORLD

Hotel

"My room has the shape of a cage
The sun extends its arm through the window
But I, who want to smoke to create mirages
Light my cigarette from the fire of the day.
I do not want to work, I want to smoke."
Guillaume Appollinaire (French poem)
Translated by R. Oropeza

Most of the information that we have obtained about the habits of pipe smoking from the 17th to the 19th centuries comes from the master painters of England, Holland, Belgium, Spain and France.(Fig. 69–72) Tobacco has been so central to the world, it should come as no surprise that the elite among the artists, poets, painters, and musicians either have used tobacco in their work or simply weighed in with their opinions pro or con. Literature is full of examples. Paintings have given us a vivid pictorial demonstration of how tobacco has had an effect on all levels of society for the past five centuries, from the very rich to the very poor.

Four giants of German literature—Johann Wolfgang Goethe (the great Napoleon of literature in the 18th century), Johann Christopher, Friedrich Schiller and Major Von Knobel—and the philosopher Kant had different views about tobacco.[130]

Fig. 69. Flemish Andrea Brower's self portrait clowning, blowing smoke through his nose. (left) He made smoking the theme of his paintings. 17th C.

Goethe expressed himself strongly against smoking. Tobacco was one of the things that he hated most. "Many ills can I bear with, and most of the things that displease me quietly bravely I meet, bowed by the will of God. Yet there are some that I hate that are worse than the

Fig. 70. Dutch painter Van Gogh (self portrait) smoking his briar pipe. Self portrait 19th C. When he was under psychiatric treatment in the hospital he stated "if tobacco is abandoned, man may live longer but not happily".

Fig. 71. La Cometa. (LL.) Majo smoking a cigarette 1855. First depiction of cigarette smoking by Goya. El Prado. Madrid, Spain. Photography by R. Oropeza. 2004

venom of serpents, four: tobacco the first, bugs then garlic." Goethe strongly believed that a well-rounded man could not smoke. He stated that smoking stupefies a man and makes him incapable of thinking. He hated tobacco so much that he would avoid going into a room where

Fig. 72. La femme de Algier by Eugène Delacroix. Note the narghile. Louvre Museum Paris 19th century. Photo by R. Oropeza. 2004

there was smoke, or if there was smoke "he excused himself that he felt unwell." Goethe made an analogy between those whose conversation he could not stand and those who smoked tobacco; both made him feel ill. He agreed with Humboldt that the "English and others had brought to Europe from America two plants, one a blessing, the potato, and the other a curse, tobacco." He was critical of those who complained, but were always idle like the "Lazy Turks" that "find a mental solace in handling a pipe and gazing at the clouds of smoke that they puff into the air. It helps them to kill time."

On the opposite side of the tobacco issue was his poet friend and adversary, the physician Friedrich von Schiller, (Fig. 73) who loved pipe smoking and snuff, though he preferred the latter like many of his contemporaries. Schiller studied medicine at the Military Academy of Stuttgart. Smoking was prohibited there, but somehow it was smuggled in by those who enjoyed it. Schiller was found with tobacco and punished by the schoolmasters, but to no avail. When he graduated, he was not only an accomplished physician and poet, but a full-blown smoker as well. Schiller loved tobacco and strongly defended its use. Schiller and Goethe,

Fig. 73. Friedrich von Schiller in Carlsbad 1791. An adversary of the Napoleon of Literature, Goethe. Schiller was for smoking—all the way. by J.C. Reinhard

the two great friends and poets, often argued the pros and cons of tobacco while in Leipzig in September 1785.

The great German philosopher Kant also disagreed with Goethe and found tobacco, taken in any form, the "best method of exciting the sensual passions." He was a snuff taker, but also smoked his pipe.

Goethe's friend Knebel, another great poet and the tutor to the Weimar princess, took a stand in the middle. Knebel agreed with Goethe about smoking and the "Lazy Turks," but disagreed with the use of snuff. Knebel criticized Goethe for disliking snuff without having tried it. (Goethe's mother was a snuff taker, which may be why he disliked it.) Knebel criticized Goethe for sniffing eau de cologne and said, "Every sensuous enjoyment has something dirty about it at the start or the finish that adds to the charm." He also wrote the following:

Though to the furthest land you fare,
And seek the hat of some outlander,
What profit? You will find it there-
Tobacco smoke and slander.

Writer Victor Hugo disliked smoking and tobacco use in any form as much as he hated the imperial Napoleon Bonaparte, who had brought the practice of cigar smoking to France during the war with England in Spain. The French philosopher Sartre, on the other hand, found smoking his best companion and the source of his inspiration. He could not think of being without it. "It tastes better when you roll it yourself . . . offers an illusion, protection against the anxiety of radical freedom."[131]

Tolstoy disliked smoking and fought against its hold all his life, without success. Sometimes he gave in to temptation and had a smoke, but only "when he had done wrong in order to deaden his conscience." Once when he killed a hare, he felt guilty, put his gun away, and had a smoke. This was a way to rationalize and justify his smoking habit. Tolstoy explained the popularity of smoking on physiological grounds. "The brain," he said, "becomes numbed by nicotine and the conscience expires."

Thomas Corneille in 1677 put into rhyme what Molière said in favor of tobacco in his comedy *Don Juan*:

Whatever Aristotle's school declare
Tobacco is divine, beyond compare
And to employ a right an idler's leisure,
No spirit is there that gives so great pleasure,
A new found medicine a heav'n-sent gain
It cleanses, brightens, and restores the brain,
Of melancholy is purgative;
Who lives without tobacco does not live.
Tobacco, Oh tobacco, that I love!

The poet and writer Robert Louis Stevenson advised women, "Never marry a Teetotaller or a man who doesn't smoke." Another great poet of the 19th century, Charles Lamb, wrote, "For thy sake, tobacco, I would do anything but die." The greatest poems written about cigarettes were published in 1880 by the poet Jules Laforgue.

Most literature in the 17th -19th centuries written against the habit of smoking was penned by reactionary, puritan conservatives of the

Christian church. Many of these writers were Jesuits. "How people can throw good money on such a folly," they argued. A German Jesuit wrote: "Money we know, is more precious than virtue, but unto them tobacco is dearer still . . . His countrymen are so infatuated that they not only smoke tobacco but actually sniff it up." "What difference," he asks, "between a smoker and a suicide except that the one takes longer to kill himself than the other? Because of his perpetual smoking, the pure oil of the lamp of life dries up and disappears and the fair flame of life itself flickers out and goes out all because of this barbarous habit."

Lord Byron, in his 1823 poem *The Island*, provided the other viewpoint:

> Sublime tobacco! Which from east to west
> Cheers the tar's labour, or the Turkman's rest;
> Which on the Moslem's ottoman divides
> His hours, and rivals opium and his brides;
> Magnificent in Stamboul, but less grand,
> Though not less loved, in Wapping or the Strand;
> Divine in hookas, glorious in a pipe,
> When tippe'd with amber, mellow, rich, and ripe;
> Like other charmers, wooing the caress,
> More dazzingly when daring in full dress;
> Yet thy true lovers more admire by far,
> Thy naked beauties—give me a cigar!

Many musicians were sympathetic to tobacco. J. S. Bach was a pipe smoker and composed a melody called *The Tobacco Pipe*. (Fig. 74) Mozart was a snuff box collector. Massenet and Fauré sang their hosannas to tobacco and cocaine. Beethoven, according to his biographer Schindler, was also a smoker, but only smoked in a tabagie, not at home. He was also a snuff taker and collected snuff boxes. The composer Franz Liszt, according to the Bettman Archives, stated that "a good Cuban cigar closes the door to the vulgarities of the world."

People in 'Show-Biz' in the 20th century enjoyed cigars or cigarettes. Smoking was often part of their performances—such as María Felix (Fig. 75), and Cantinflas (Fig. 76), two of the most famous of the Mexican entertainment world.

Fig. 74. The Tobacco Pipe was a composition by Johann Sebastian Bach. He also composed a sonata to coffee. Other musicians who used tobacco were Beethoven, who smoked a pipe, Wagner, who inhaled snuff, Sibelius, who smoked cigars and Stravinsky, who smoked cigarettes. From Anna Magdalena's Bach's music-book (Prussian State Library, Berlin)

Fig. 75. María Félix Mujer. (María Bonita) a Mexican actress and icon of the last half of the 20th C. She smoked all her life and died in her 80's. Photograpy by Edmundo Báez 1943

Fig. 76. Cantinflas. He became known to the American public in his film with David Niven, Around the World in 80 Days. A Mexican comedic entertainer during the last half of the 20th C., the cigarette was integral to the last part of his act. His punch line was "Ahí está el detalle." (That is the question.) Riddles of Cantinflas.

17.

WOMEN, CHILDREN, AND TOBACCO

"She took it like a mouthful
of spring water on a parching day."
Jack Yeats, brother of the poet William Butler Yeats, speaking of a woman taking snuff in Scotland.

"They all smoke, men and women, and a pipe
an inch large serves the whole family several
years, and though never so black or fowl is
never suffered to be burnt. Seven or eight will
gather to the smoking of a pipe, and each taking
two or three whiffs gives it to his neighbor,
commonly holding his mouth full of smoke
till the pipe comes about to him again."
G. O'Brien

Smoking, chewing, or snorting tobacco in most pre-Columbian cultures was mainly the domain of men. Most priests and shamans were males, but some women were shaman-herbalists ("La Curandera"). Women still take care of the sick in the towns of Latin America and use tobacco as part of their "medical armamentarium or 'power tools'." In some cultures of North America, women also smoked, but they were never allowed to smoke the Calumet Pipe. That was a sacred pipe of war or peace to be smoked by men only. Women and children of the Orinoco Amazon Basin chewed (sucked) green tobacco wads, as described by Vespucci five hundred years ago, and they still do today.

During the beginning of the 16th century most European smokers were sailors, who propagated this habit to people around the world. Male landowners and slaves in the Spanish colonies smoked, but few Spanish women did, probably because it was not accepted in the conservative Spanish court of Ferdinand and Isabella. The practices of snuff taking and pipe smoking were also taken up by Queen Catherine (Catarina) de Medici at her court in France at the end of the 16th century, with the pretext that tobacco, as recommended by Nicot, would cure her migraines. After she had broken the ice, all women and men at the court smoked, and the rest of the country followed. During the 17th century, after Raleigh's influence with the aristocracy and the English court, smoking became accepted by men and women at the court and was considered part of being chic. Queen Elizabeth smoked a pipe under Raleigh's influence. Some women smoked slender cigars in long cigar holders.

In other countries, such as Holland and Germany, women took to tobacco with gusto. As described by the 18th-century poet Carlyle, "Women young and old were seen smoking at the table." Chinese and Japanese women smoking long pipes are depicted in paintings in spite of a prohibition against smoking. Pipe smoking was never very popular with European women; it was mainly a practice of men but women shared tobacco with men, especially, when snuff-taking became the mode of tobacco use in the 18th century. But some smoked pipes (Figs 77–78) and cigars. (Fig. 79)

When cigar smoking became a practice at the end of the 18th century and the beginning of the 19th century, women like George Sand smoked cigars but most preferred snuff. Some women became tobacco chewers, but when cigarettes came into fashion they took to it with enthusiasm. Smoking cigarettes became elegant and acceptable in society. Some women smoked cigarettes during the first quarter of the 20th century when they first became enfranchised, but smoking still remained primarily a man's habit. Men were even encouraged by the United States government to smoke. Tobacco was part of the army ration as early as the Civil War, and this practice continued afterwards during the wars of the 20th and 21st centuries.

During World War II women became liberated. When five million women took the place of men at the war factories and at almost every other workplace, they wanted to be equal to men in every respect. Women took on men's work in the factories, wore pants, and became ardent smokers from then on to the present time.

Fig. 77. Woman smoking a pipe in a Dutch tabagie. From the beginning, women shared tobacco with men. Painting by David Teniers (Antwerp 1610–1690) Paul Granpe's Collection.

"People snuff so often that their noses are more like a dust-heap than a nose."
Johann Heinride Cohausen, 17th century

Women were heavy users of snuff in the 17th—18th centuries. French aristocratic women had to know the rituals of this practice, because during important gatherings at court they had to take snuff with grace. It was well accepted in society, and it was chic to sneeze after snuff taking. It was also acceptable to clean their noses with a fanciful but dirty handkerchief that was readily available on their sleeves. Men apparently did not find it disgusting to see their women with yellow noses (maybe because their noses were tainted as well).

A British observer watched a young woman smoking "a long jessamine pipe, which, however, most Eastern ladies do."

As for children, there were no laws prohibiting them from smoking until the 20th century. (Fig. 80) Then, in 1908, an English law prohibited

Fig. 78. Madam Vigée Lebrun (1755–1842) Self portrait. She was the official painter of Marie Antoinette at Versailles. Smoking dainty pipes was chic and fashionable at the court of Louis XVI. Dunhill Collection. London.

selling tobacco to children and banned adolescents from smoking in public places. At the 20th century's end, there was increasing concern about teen smoking, and laws were passed to try to curtail it.

"Most Irish families subsisted chiefly on potatoes and sour milk," wrote John Sterman. "They all smoke, women as well as men, and a pipe an inch long serves the family several years and though never so black or foul is never suffered to be burnt." Samuel Rowlands in 1628 gave an unflattering picture of a merchant's wife "living above her situation." She was unhappy "except (contrary to the modesty of her sexe) shee be halfe of the man's fashion: she jets, she cuts, shee smokes, she drinkes, and what not that is evill."

On Yap Island in the Pacific, an adventurer found that "women of higher rank had perfect luxuriant figures, and like their men were . . . passionate smokers. It almost seems as though the feminine sex, especially the young girls, are possessed by this habit learned from the Europeans. A cigarette is seldom absent from the lips of the beautiful and graceful smoker."[132] Tobacco was also smoked in pipes in the South Pacific Islands because it was introduced to them in that form by Captain James Cook.

Fig. 79. La Lionne. Cigarette smoking lady. Printed in Paris by Othan. Published in London by Gambart (1846)

At the present time more than ten million American women follow the cult of the Beautiful Nicotiana, and they are dying of cancer of the lung and other tobacco-related diseases just like men.

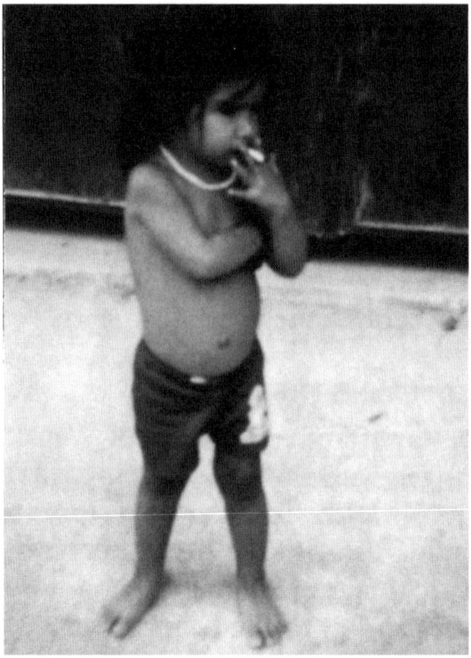

Fig. 80. A poor Mexican child smoking a cigarette butt probably found in a 'basurero' (garbage dump). Cover of the WHO publication.

18.

WALKING A MILE FOR A CAMEL

The branding of tobacco products in the 18th century represents the first form of tobacco advertising. The first of these was the Lorilard Company in 1760.[133] The 17th and 18th century trade cards and wrappers were identifiable by symbols of the brand, such as the Indian head or tobacco leaves. Before 1876–1888 the four big cigarette companies had advertised, but it was the marketing of Camels, introduced in the United States in 1913, that gave birth to modern advertising. Signatures of Camels' advertising were the use of an identifying slogan ("I'd walk a mile for a Camel"), the advertising on the package itself, and free giveaways. (Fig. 81)

Other cigarette advertising trends included the targeting of specific audiences such as women. (Fig. 82) For males there was the Marlboro Man's appeal to masculine virility, which first appeared in the United States in the 1920s and has remained as an advertising trademark ever since. Other advertising techniques appealed to vanity and health concerns. Ads promoting smoking as a substitute for eating sweets were linked to the emergence of the slim, modern American woman. Now men and women are overweight and McDonald's has come under attack, giving a respite to the big tobacco companies.

To forge future converts in early childhood, images of Mickey Mouse were stamped on packs of cigarettes and on boxes of cigars. Other gimmicks were to target specific audiences through the development and marketing of specialty brands, such as Kools (the first mentholated cigarette), Viceroy (the first filter-tipped cigarette), and Pall Mall (a king-sized cigarette.

Fig. 81. Advertisement for cigarettes. Gas station Route 4, New Jersey, 2004. Photography by R. Oropeza (2005)

Fig. 82. Marlboro advertisement from a poster. Mother and child, 1930's. Photography by R. Oropeza (2004)

Cigarettes were at first considered effeminate by those who smoked cigars at the end of the 19th century. With the success of cigarette advertising, they came to be considered elegant and masculine for men, and feminine for women. Walking a mile for a Camel and the Marlboro Man succeeded, and the world would never be the same again.

An epitaph to a smoker
"Here lieth he who had lived longer if

He had not choked himself with
Tobacco whiff."
Unknown author

"The light of his day had
already been extinguished
when his pipe was still lit."
Grabner, 18th century

"If my eyes want to
Follow your smile
I see that I am to
End like you."
J.V. Gijsen, 18th century

Tobacco is such a disease that ruins the body
And as medicine, it avails little against death.
It does more evil that good. Whoever you are
I beg of you: deliver us from this poison and this plague.
Petrus Scribanus, 1628

"[Tobacco] . . . that as it will ever be
condemned, it will ever be retained."
F. Hoffman, 1746
(A correct prophecy)

19.

TOBACCO, THE WORLD PLAGUE OF THE 20TH—21ST CENTURIES

CIGARETTES

Man's Best and Worst Friend. The Ultimate Triumph of Tobacco

The 20th century was the most exciting century yet for humankind. It was the age of high technology and progress that transformed the fabric of society. The world would never be the same. Man lived in a faster, more complicated society where stress had to have an escape valve to release the frustrations of daily life. What an easier way than to have at hand a pacifier to make life more bearable? The cigarette and the cigar were the answers for many people. Tobacco was a "friend" in moments of stress, frustration, loneliness, or boredom, a companion for those who looked for creativity or inspiration. Tobacco had been maligned for many reasons by millions of people for centuries, but for many others it played a seemingly positive role in their lives; it has become the most convenient pastime. We have to look at both sides of the coin. (Figs. C38, 83–84)

After World War I cigarettes became the most popular form of tobacco consumption around the world. The tobacco used for cigarettes was scented and soft to the palate. The cigarette was introduced in the United States in the last quarter of the 19th century by the upper classes that had spent time in England and had taken to cigarette smoking: "Some of the ladies of this refined and fashion forming metropolis [New York] are aping the silly ways of some pseudo-accomplished foreigners, in smoking tobacco through a weaker and more feminine article, which

Fig. 83. Old Man of Venice at Piazza San Marco (St. Mark's Plaza). He enjoyed the cigarette for life, a faithful friend that never let him down. Photography by R. Oropeza (2004).

has been most delicately denominated 'cigarette.' " After a few years this habit was well accepted in the United States and some moralists despaired, "when one hears of sly cigarettes between feminine lips at croquet parties."

Cigarette smoking started in the 18th century, but cigarettes as we know them now started in the middle of the 19th century. As far back as the 8th—15th centuries, however, the Maya minced tobacco and rolled it in a leaf of *zapotill*e (custard applie) or in a thin corn husk. In the middle of the 19th century, during the industrial revolution, cigar or pipe smoking became inconvenient and the cigarette was the answer. It was easy to light with a simple strike, easy to keep in the pocket or purse, and ready to be smoked at any time and shared with a good pal.

According to Rivat's book, *Cigarettes are Sublime*, the cigarette arrived in France around the years 1825–1830, having come from Spain. They called it a *cigaritto*, "a little square of letter paper into which one inserts tobacco." Casanova after his travels in South America in the 19th century claimed to have introduced cigarettes to Europe. He described cigarettes in his memoirs and called them "paper cigars." They were made by the Austrian Regie Factory.[134]

During the Crimean War cigarette smoking became widespread. It was actively encouraged by both sides of the conflict—Russia on one side, and England and Italy on the other.[135] It was soothing to smoke while in battle or when wounded and lying at the filthy barracks of Scutari Üskutar, Constantinople (Instanbul), across the Bosporous Strait. (Fig. C39) A similar phenomenon had occurred during the Thirty Years War in Europe in the 17th century, when pipe smoking spread rapidly throughout Europe and the rest of the world. Wars have always been the catalyst for the spread of tobacco in any form.

Louis Napoleon III, Emperor of France and an ardent smoker in the mid-19th century, found cigarette smoking delightful. Cigarettes became a mainstay at his court.[136] During the battle of Solferino in 1859, when Louis Napoleon III helped the Piedmontese win Milan and Lombardy from the Austrians, he kept muttering to himself. "poor creatures, poor creatures, what a terrible thing is war," all the while clenching and puffing on his cigarette or his cigar.

The Processing of the Leaf

The tobacco for the filler leaf is selected and grown in special shaded places. It is then cut and piled in stacks to ferment. A thin silk leaf is wrapped around the core of the cigar or filler. In the United States one of the best wrappers is grown in the state of Connecticut.

For about two centuries, harvested leaves were carted to curing barns to be dried out by the open-fire method. Then, in the early 19th century, the safer flue-curing process gradually replaced fire-curing. Flue-curing "brightened" the color of the tobacco leaf, which was grown mainly in Virginia and North Carolina and was distinguished by its mild taste. This "bright tobacco" by the early 20th century constituted about 50 percent of the content of cigarettes. In contrast, Burley tobacco, grown mainly in Kentucky and Tennessee, was used for cigars and snuff. Two other methods of processing were used—air and sun curing, the methods used by the aborigines of America. Until the second decade of the 19th century, tobacco manufacture was characterized by many small factories employing relatively few workers in comparison to other, contemporary industries. Then technological advances and changing consumption patterns, among other things, led to the modernization of tobacco manufacturing. In the last half of the 19th century, hand-rolled cigarettes (called *cigarritos* in Spain) became popular in Europe.

In England, Phillip Morris began selling different brands to customers. At the beginning they were produced in a tedious process and in small amounts by tobacco factory workers, but the invention in 1885 of a practical, cigarette making machine made possible mass production. In 1869 the Kinney Company manufactured the first mass production cigarettes. The cigarette had a tip that prevented the skin of the lip from being in direct contact with the tobacco, and thus prevented the yellowish-brown discoloration of the skin. Another manufacturer that became very successful was Virginia's W. Duke, Sons and Co., which by 1890 was the number one producer of cigarettes in the world. This led to the subsequent rise in the United States of the tobacco trust. Until 1911 (the year that the government broke the trust up into four new companies) the trust exercised monopolistic control of tobacco production and sales (but not cultivation).

The organization of multinational tobacco companies and expansion into overseas markets was already underway by the early 20th century. China and Japan were huge importers of United States cigarettes. India and Africa opened up soon afterwards. Now China is the main producer and exporter of tobacco.

Although Robert Boyle in 1680 had invented a mixture of sulphur and phosphorous that ignited with a light friction, it was Kammerev Ludwigsburg, a chemist, who invented a perfect match, one that could be kept in the pocket without igniting and catching fire with slight friction. Austria had also developed at about the same time, in 1865, a good quality cigarette. This was the beginning of widespread cigarette smoking in Europe, and soon after in America. The first cigarettes were called "Turkish," because the tobacco was imported from Turkey. American entrepreneurship in mass advertisement resulted in a much higher consumption in America in a short period of time. This mass cigarette advertising became the model for automobile and other industrial or commercial products.

Mass target advertisement, appealing to different groups of people, would be an important factor in converting new tobacco lovers. This advertising concept came at a time when the radio was universally used throughout the world. The American government, by having cigarettes as part of the rations of the army during the first and second world wars, sent thousands of new cigarette ambassadors to Europe and Asia, in this way making "new converts to the cause." Then came television advertising for different brands of cigarettes appealing to different groups of society, until cigarette advertisements on television were banned in America in the 1960s. This would hardly put a stop to smoking. By that time it was a part of the American culture. Adolescents took up cigarette smoking as a rite of passage to adulthood. It made them feel important, grown up, and accepted by their peers. It was not unlike the ritual of passage for adolescents that was practiced by the Maya more than one thousand years ago. Tobacco had been challenged again, and the nicotiana had again proven to be resilient and triumphant.

The cigarette became a symbol of sublime enjoyment to be shared with friends and co-workers on a short coffee/smoke break. It is an elegant ritual to share a cigarette with a friend, opening the pack, tapping the cigarette after it has been removed from the pack, offering a cigarette to the friend, lighting it for him, then after first taking a good look at the cigarette and giving two or three deep puffs to get the needed nicotine rush, tapping the ashes quickly, and finally engaging in conversation. It was yet another in a long line of tobacco rituals such as smoking the Calumet on special occasions or the ritual of taking snuff.

Cancer of the Lung, and Other Tobacco-Related Diseases

Our knowledge and understanding of what cancer is dates back only about 150 years; therefore, all that has been said about cancer in the previous two thousand years becomes entirely anecdotal. Included as cancerous tumors in those years were all kinds of inflammatory and other masses that were interpreted as cancer without having a histological diagnosis.

The plausible or possible role of tobacco smoking as a cause of or as an important factor in some cancers, in particular cancer of the lung, oral cavity, larynx, and bladder, and as factor in cardiovascular disease, wasn't suspected until the beginning of the 20th century. It was not epidemiologically and experimentally proven until the middle of the 20th century. To give an example of previous suppositions: syphilis was once considered one of the main causes of cancer of the mouth.

Frederick Hoffman in 1746 in his treatise, *An Epidemiologist, Diseases of Tradesman*, gives an account of the diseases of tobacconists and those who took snuff. Hoffman vaguely made the connection between tobacco and lung disease: "that affects the lung . . . produces a narcotic effect upon them . . . dry the tender coat of the aspera arteria [coronary artery]." Then he states, "I have known many brought to consumption by chewing tobacco . . . so bewitching is that incurable custom of chewing and smoking tobacco . . . That as it will ever be condemned, it will ever be retained." (In this he was a prophet.) This was the sort of inaccurate and vague connection that was made between tobacco and cancer or other diseases in the 18th and 19th centuries.[137]

Anecdotes of two patients with tobacco-related cancers were published by Dr. John Hill in 1761. He describes two patients, both snuff takers, with tumors of the nasal passages. On the first case: "an unfortunate gentleman who after a long and immoderate use of snuff, perceived that he breathed with difficulty through one of his nostrils. The complaint gradually increased until he perceived swelling (within his nose) . . . it grew slowly till the end. It filled out the whole nostril, and swelled the nose so as to obstruct the breathing. He found it necessary to apply for assistance. The swelling was quite black [I suspect it was a melanoma rather than a carcinoma]. An adherence by a broad base so that it was impossible to get it away . . . and the consequence was the discharge of a thick sharp humor with dreadful pain and all frightful symptoms of cancer . . . and he seemed without hope when I last saw him." The second case was a woman with similar symptoms who also was a snuff taker. She had a tumor of the nostrils that followed a similar course as with the first patient. (This most likely was a carcinoma.)[138]

Lancet, the prestigious British medical magazine, in 1857 mentioned that tobacco "acts by causing consumption, hemoptysis and inflammatory conditions of the mucus membrane of the larynx, trachea and bronchus." There were no x-rays available at that time, so it is impossible to know if the patients referred to had cancer of the lung, tuberculosis, or something else. This was the poor state of knowledge related to tobacco and cancer one hundred and fifty years ago.

In 1859, Bouison reported sixty-eight patients in France with cancer and he found that only one patient was not a tobacco abuser. This seems inaccurate and highly exaggerated, but probably almost everybody was smoking at that time. One of those patients had cancer of the lip where he rested his pipe.[139]

The father of surgery, Theodore Christian Billroth (Germany), (end of 19th c.) and William Halsted (19th—20th century), the father of surgery in the United States and one of the most influential surgeons in the 19th and 20th c., as well as Dr. William Osler (end of the 19th century), an icon of medicine in the United States and throughout the world who worked with Halsted at Johns Hopkins—none made any mention of tobacco as a factor in any cancer or any other diseases related to tobacco.

Billroth never mentioned any of his patients' cancers being related to tobacco, even in those who were smokers with cancer of the larynx (Billroth removed the first larynx with cancer in 1872). Billroth disliked smoking, and in 1889 he commented, "Smoking is a disgusting vice injurious to the health." He said that "society is becoming more and more neurotic and this is due to alcohol and tobacco." But he did not mention in which ways tobacco was injurious. His assistant, Von Bergman, was involved in an international dispute with his counterpart, MacKenzie, of England, at the end of the 19th century while treating Prince Frederick III of Germany for cancer of the larynx. MacKenzie insisted that the tumor was of a benign nature, even when the then-king was dying of cancer. Frederick III was a heavy smoker, but no mention was ever made that this cancer was possibly the result of his smoking, in spite of the fact that he would smoke in front of them just before his death until his voice could barely be heard.

Two of our United States presidents, Ulysses Grant and Grover Cleveland, both cigar smokers, developed tobacco-related cancers (larynx and oral cavity, respectively), but tobacco was never mentioned as a causative factor. Was it not obvious, as it should have been, that smoking had something to do with their cancers? In the meantime the United States government was supplying tobacco as part of army rations from the American Civil War up to the wars of the 20th-21th centuries. General Pershing, the American commander-in-chief in Europe during World War I, wrote in a letter to Washington: "Tobacco is as indispensable as the daily ration. We must have thousands of tons without delay."[140]

In 1880 an antitobacconist journal included an article by Dr. Conan Doyle of Plymouth, England, connecting clay pipe smoking and cancer of the mouth or nose: "went up county once, and operated upon an old fellow's nose which had contracted cancer through his holding the bowl of a short clay pipe immediately beneath it. I left him with an aristocratic, not to say supercilious organ, which was the wonder of the village."[141]

In 1930 a German scientist named Morten wrote an article entitled "Possible Cause of Cancer of the Lungs." This is probably one of the first clinical, scientific observations connecting tobacco and cancer of the lung. There were other observations, however, that most patients with cancer of the lung were cigarette smokers. Alton Ochner and the famous (and still alive at more than ninety years of age) Michael De Bakey in 1934, 1939 and 1941 made the observation that 95 percent of their patients with cancer of the lung were cigarette smokers (this is a very accurate statement). Ochner stated, "Man is the only animal that pollutes his body structure by smoking and nature is teaching him a lesson."

Heavy cigarette smoking, as we know it now, only started at the beginning of the 20th century when cigarettes were mass produced and packed and sold successfully by mass advertisement. About 30 years later, it was the right time for cancer of the lung to manifest itself.

Ochner made the sarcastic remark that "the first thing that a child learns from his mother's lap today is to watch out for the hot cigarette ashes from her burning fag." When Ochner was asked in 1954 why women smokers did not have the same incidence of mortality for cancer of the lung as men smokers (cancer of the lung in women at this point was already on the rise, however) he was quick to answer, "Give them time and they will." He was right on the money. By the 1960s, cancer of the lung in females had increased six fold.[142] By the end of the 20th century, the incidence and mortality of cancer of the lung in women had almost caught up with that of men.

The sharp rise in cancer of the lung in women started in the 1960s at the time that, coincidentally, women were demanding equal rights with men. They got what they asked for: equal rights to work, and equal rights to get cancer of the lung since they were equally smoking cigarettes for the past 30 years. It takes about 30 years from the time one starts smoking to develop cancer of the lung, and less time if smoking starts in adolescence.

Under the direction of one of our icons of medicine in the United States, surgeon Evers Graham, who himself was a smoker and died of cancer of the lung, the first successful removal of cancer of the lung was performed in 1933 on a Maryland physician smoker (who survived Dr. Graham). Dr. Graham and Dr. Ernst Wynder, an epidemiologist who later worked at Memorial Sloan Kettering Cancer Center in New York, proved statistically, in control studies of smokers vs. non-smokers, the causal relationship between smoking and cancer of the lung. At the same time another epidemiologist working in England, Dr. Richard Doll, came to the same conclusions. This was the first scientific link and became the conclusive turning point. This control study of 684 smokers was published in the JAMA 1950:143–329–36 and entitled "Tobacco Smoking as a Possible Etiologic Factor in Bronchogenic Carcinoma."

Dr. Oscar Auerbach in 1957 conducted experimental studies in beagles that were forced to smoke through a tracheostomy (opening in the wind pipe). It was a cruel act, but this experiment gave us valuable information when post mortem examinations of these animals that were subjected to smoking proved that there were pre-cancerous changes in the lining of the lungs (epithelium). Dr. Oscar Auerbach also did extensive microscopic studies of lung tissue in smokers versus non-smokers. He reviewed ten thousand slides of tissue samples for this project.[143] This was another important landmark in proving the causal relationship between tobacco and cancer of the lung.[144]

From the middle to late 20th century, many famous people started dying of cancer of the lung or larynx, such as King George VI, Babe Ruth (larynx), Humphrey Bogart (larynx), Yul Brenner (lung), Leonard Bernstein (lung), and mafia boss John Gotti (larynx). Communities throughout the United States in the late 1960s became alarmed and fully aware that smoking was a health hazard and something had to be done about it.

After thirty years of denial that such a causal relationship existed between smoking and cancer of the lung, and further denials that nicotine was addictive, and after spending billions of dollars lobbying in congress against antismokers, tobacco companies finally accepted the facts. They first tried diversionary tactics such as inventing new cigarettes with less nicotine, a better filter, shorter cigarettes, etc., and also fought lawsuits, often losing in the courts. More recently (2003), Phillip Morris went on television offering advice on how to quit or prevent smoking.

The Present State of Affairs

The American Cancer Society published in 1954 the results of the research by E. Graham and Ernst Wynder that definitely concluded that there was a causal relationship between smoking and lung cancer. By 1960 researchers had found that the tar and nicotine in cigarettes, when inhaled, could cause not only cancer of the lung, larynx and oral cavity, but also heart disease and peripheral vascular disease (hardening of the arteries). Such findings initiated a societal shift in the perception of cigarettes that has continued until today.

In the United States anticigarette momentum put pressure on the U.S. Congress in the 1960s to pass legislation requiring manufacturers to add a warning label on cigarette packs and cartons. Cigarettes could no longer be advertised on radio or television, and manufacturers had to place a health warning in cigarette ads in the print media. Joseph A. Califano Jr., a lawyer and reformed smoker, was influential in bringing in these changes while in charge of the Department of Health HEW in 1978 and while in OSHA in 1979. The war on cancer by President Nixon in 1971 was supposed to conquer cancer by the end of the 20th century. It was a crash program to rival the race to the moon in the previous decade by President Kennedy. The space program was successful and the moon was reached, but the conquest of cancer was not.

Pumping millions of dollars into basic research has without a doubt accomplished much in the area of a better understanding of the behavior of the "outlaw" cancer cell. We know much more about the molecular and genetic changes that occur in these cells from the effect of carcinogens, but we have not made much progress in prevention of some cancers such as cancer of the lung. This cancer could be practically eradicated simply by changing the behavior of society in relation to smoking. Many of us make a living either from tobacco cultivation, cigarette production, cigarette sales, cancer research programs, or treating patients who smoke and consequently develop cancers; however, we should not accept or justify the practice of smoking.

In the 1990s the tobacco companies faced escalating numbers of individual and class lawsuits against them based on the health hazards of cigarettes. Also in the 1990s and at the beginning of the 21st century, stringent restrictions on smoking in public buildings, airplanes, restaurants, and elsewhere were enforced to placate the antitobacco groups that not only dislike smoking, but are afraid of getting cancer of the lung from second-hand smoke (which accounts for about 5 percent of cancers of the lung). More recently, in New York City in 2003, another "sin tax" was added to the 1990s Clinton tax "to discourage the good citizens of New York from smoking." (Mayor Bloomberg)

The war between antitobacconists and tobacco proponents still goes on after five hundred years. In previous centuries the church tried to modify tobacco use. Now it is the state, through public health programs and other measures such as banning advertisement and funding educational programs in schools and in other institutions. There are also private groups that are against smoking in public places, not because it is harmful to their health, but because they just don't like it.

The incidence of cancer of the lung in the United States (165,000 new cases per year) has changed little, and the mortality rate is still very high: 85 percent in the United States, and 95 percent in the ghetto of Harlem, New York. With no satisfactory methods for early detection, most patients come in for treatment with advanced cancers, for which the modalities for cure or palliation with radiation or chemotherapy are very poor. Treatments offer little extension of life and practically no improvement in the quality of life.

The American Cancer Society has a track record of fifty years of spending millions and millions of dollars on research and on anti-smoking campaigns at the national and community levels. All approved cancer programs in the United States must have specific programs directed to changing the smoking habits of the communities. The anti-smoking efforts have resulted in modest gains in the United States but not in the rest of the world.

The largest medical and oncological society (ASCO) in the United States, with 20,000 members, held a meeting in Chicago in May 2003. The president of the society gave a wonderful oration wherein he made a call and a pledge for a "tobacco free world" (*déjà vu* all over again). But he stopped short mentioning what he had in mind: to change the habit of smoking that has already been tried, and failed, over the past five hundred years. At least the president of this society acknowledged that a similar call had been made in 1996 "with little effect."

Tobacco has faced banishment from the face of the earth for five centuries by the most powerful popes and muftis, who considered tobacco use in any form as unchristian or against the Koran. Tobacco use has been condemned by kings and other potentates because it was contrary to good moral or social habits. More recently, smoking has been assailed because it has been proven to be hazardous to health. Still, tobacco has not been banished from the face of the earth in spite of the misery that it has brought to many people. No one is against taxing tobacco, however, which may be a good part of the problem.

Tobacco was one of the main foundations on which the great nation of the United States of America was built—at a high cost in suffering and death, beginning with the many thousands of slaves who toiled at the plantations. On the other hand, tobacco smoking has meant a lot to millions of people. It has brought inspiration and comfort to many, brought riches to governments through taxes, and reaped profits for companies and corporations, provided a source of employment for many, consoled the spirit of the sad and desperate, and given soothing enjoyment and comfort to millions in peace or war. In spite of this we should not condone its use.

The pipe says Thackeray "draws wisdom from the lips of the philosopher and shuts up the mouth of the foolish. It generates a style of conversation, contemplative, thoughtful, benevolent and unaffected. May I die if I abuse that kindly weed which has given me such pleasure!"

For those who enjoyed tobacco and paid the high price of cancer or cardiovascular disease, or emphysema, we must feel sorry; however, they knowingly spent their lives and money enjoying the weed. For those who are not yet initiated into the lures of tobacco, we must try to prevent them from taking up the habit by continuing educational programs that stress the dangers and consequences. Only through education during childhood and before the rite of passage of adolescence might we prevent this most damaging habit from spreading. Once an adolescent has been initiated into the practice of smoking, it is almost impossible to correct the habit—the addiction. The addiction is more powerful than the desire to quit.

I am afraid that tobacco is here to stay for a long time to come.

20.

MILLENNIUM LEAF
The Future of Tobacco

"To stop smoking is hard as
to make him stop walking."
Alfred Lord Tennyson. (He always had a clay pipe at his side)

At the beginning of this millennium. it is predicted that the targeting by cigarette manufacturers of third world markets will result in what may be termed a "global epidemic." The truth is, it already is an epidemic. More than one million people die annually of cancer of the lung throughout the world, and many millions more die of other tobacco-related diseases. Only the plague of the 15th and 17th centuries and the AIDS epidemic of the 20th century have killed more people in one year than tobacco.

The industrialized countries will probably see more drastic legislation coupled with more restrictions for smokers on where they can or cannot smoke. Legislatures will pass more and more unsuccessful measures and laws. In the past however, governments have always benefited from the tobacco industry. The "sin taxes" increase their revenues. This is punishment, not prevention, and creates a conflict of interest.

None of the drastic measures of the past has discouraged smoking. Those who love smoking take these prevention programs, mandates, and restrictions as a gauntlet thrown down by the antismokers. There are twenty-five million American smokers, and many more millions around the world to be reckoned with by antitobacconist groups. The tobacco lovers feel that the government and society are infringing on their individual rights, and that they should have the prerogative to decide for

Fig. 84. Neopolitan girl, a passionate smoker of cigarettes. Cigarettes were seldom seen absent from her lips. Closing the old generation gap. Photography by R. Oropeza. 2004

themselves what is good for their own lives and bodies. Tobacco users have always organized their own smoking clubs where their habit can be enjoyed without disapproval or censure.

Smoking is also promoted through magazines such as Cigar. The cigar has again emerged as a status symbol. It has begun to percolate and may gain momentum in response to the anti-cigarette smoking movement. There is not as fierce a frenzy against smokers of cigars as there is with cigarette smokers.

There are clearly conflicting motives for government involvement in tobacco. Especially in the United States, growers lobby strongly for their industry, citing employment and taxes as important factors in their survival. In the legal wrangling over whether or not the tobacco industry should be held responsible for the health costs of tobacco users, some retorts have sounded too cruel to be believed. For instance, one argument is that the government will save money in terms of Social Security benefits for those whose death comes early due to cancer of the lung or other tobacco-related diseases. It is a fact that smokers shorten their lives by several years, even if they do not develop cancer. The oppossing argument pleads that the decline in quality of life and the costs of health care for smokers and their families are the factors that should influence our policies.

From the history of tobacco over the last five hundred or so years, the conclusion is clear: Because of the highly addictive power that nicotine has on human beings, once the addiction to tobacco has been established it is very difficult to stop.

The solution is not so clear. Perhaps our only alternative is to teach our children not to start smoking, and hope that continued and emphatic education will prevent future generations from use of this "devil stick."

There remain two sides to the story.

When a tobacco smoker wakes up in the morning, he/she thinks or sings, "What better way to start the day than with the kick that tobacco and coffee give in the morning!"

'At that blest hour when nicotine and coffe (coffee),
The day's first cup and cigarette, unite
And soar into the brain with sleep still fluffy
Like bee and queen be (bee) on their nuptial flight.'

The Two Commissars, Bombay 1944

On the contrary the non-smoker will wake up whinning in the morning:

"Filth of the mouth and fog of the mind . . .
Hemlot, aconite—(tobacco)

Charles lamb 1909 London.

Cantinflas, the best known actor-comedian and an icon in México, made it a part of his act to puff on a cigarette while murmuring these words at the end of his comedy: "ahí está el detalle" (that is the question). To smoke or not to smoke. (Fig. 76)

Fig. 85. Cantinflas. Ahi está el detalle (That is the question). To smoke or not to smoke.
Engraving by a Maya of Yucutan. Photography and private collection of R. Oropeza.

ENDNOTES

Introduction

[1] Ernst, 133–41; Mason, 1.
[2] Ernst, 133–41; Mason, 1.
[3] Spinden, 33.
[4] Spinden, 32.
[5] Porter, 152.
[6] Porter, 141; Spinden, 23–24, 28–33.
[7] Spinden, 29–30.
[8] Rosenzweig, 39.
[9] Hernández, :24–25.
[10] Porter, 165.
[11] Schondube, personal reference.
[12] Porter, 142.
[13] Robicsek, 1–3.
[14] Heimann, 24–25.

Part I

[1] Fernando Morbán, personal communication, Museo del Hombre, Dominicano Santo Domingo, Dominican Republic.
[2] Cárdenas, 150; Porter, 50
[3] W. A. Penn, 11, 1905.
[4] W.C. MacLeod, 1930; Robicsek, 21; Furst P.T., 3–54.
[5] Bezoni.
[6] Robicsek, 19–20; Porter.
[7] Robicsek, 11–20.
[8] Schondube, Curator of the Museum of Anthropology, Guadalajara, México, personal communication.
[9] Robicsek, 21–22.
[10] Peter Furst, 1974.
[11] By La Barre, Coe and Wilbert (Furst's book).
[12] Aguírre, 13–16, 128; J. Morton.
[13] Museo del Hombre Santo Domingo.
[14] D. Silver, 1966.
[15] Svi Dor-Ner, 1–5; S. Morison, *Admiral of the Ocean Sea,* 22.
[16] Svi Dor-Ner, 50–67.
[17] Dyson, 41–56.
[18] Cohen (Rosenzweig's book).
[19] Svi Dor-Ner, 6–16, 17–23; Dyson, 23.

[20] Svi Dor-Ner, 32.
[21] Dyson, 68–69; S. E. Morison, *AOS*, 30–32.
[22] Dyson, 80–81.
[23] Riva Palacio, XXXIII.
[24] R. Fuson
[25] Dyson, 99.
[26] Fuson, 75, 82.
[27] Navarrette (1890); Heiman, 6.
[28] S. Morison, *Admiral of the Ocean Sea*, 69.
[29] Morison, *AOS*, 70–93.
[30] Morison, *AOS*, 98.
[31] Fuson, 100; Las Casas, 227.
[32] Morison, *Admiral of the Ocean Sea.*
[33] Morison, *AOS*, 260; Fuson, 103; Las Casas, 227–230.
[34] Las Casas, 229–231.
[35] Robicsek, 5.
[36] Fuson, 137–151.
[37] Fuson, 158.
[38] Fuson, 158.
[39] Las Casas, 290.
[40] Fuson, 160–61; Las Casas, 290–93.
[41] Fuson, 161.
[42] Published in 1535.
[43] Brooks, 15.
[44] Brooks, 15.
[45] Spinden, 27–28.
[46] Corti, 51; Spinden, 30.
[47] Spinden, 37–39.
[48] Spinden, 31.
[49] Spinden, 23–24.
[50] Spinden, 32–33.
[51] Spinden, 34.
[52] Morison, 44.
[53] Porter, 139.
[54] Porter, 139.
[55] Moya Pons, *The Slave Trade of the African Continent*, 28.
[56] Heimann, 25.
[57] Moya Pons, 45–47; *American Heritage History of U.S.* Vol. 1, 59–76.
[58] Moya Pons, 91–92; Las Casas, 585.
[59] Besson, 167–171.
[60] Moya Pons, 91–92.
[61] Corti, 51.
[62] Spinden, 25, 155–156.
[63] José Alcína Franch, Museo del Hombre Dominicano, 107.
[64] Gonzalo Fernando de Oviedo y Valdéz, 1514–1526, cited in Corti, 41–42.
[65] Morbán, Museo del Hombre Dominicano, Santo Domingo, personal communication; Arevalo, 1–56.
[66] Casa, 1977 b.
[67] José Alcína Franch, 109–115.
[68] Fernando Morbán Laucer, Director, Museo del Hombre, Santo Domingo, personal communication.
[69] Spinden, 54–57.
[70] Museo del Hombre Dominicano Santo Domingo, D. R.
[71] Las Casas, 275.
[72] Peguro, 157–175; Esteban-Deive, 1–428; Lizardo, 1–332.
[73] Heimann, 14–15.
[74] Spinden, 23–24.
[75] Vega, 3–14.
[76] Corti, 41–44.
[77] Brook-116.
[78] Corti, 41–42; Heimann, 24.
[79] Spinden, 105.
[80] Spinden, 101.

[81] Las Casas, 290–92; Morison, *AOS*, 423–24.
[82] Robicsek, 116.
[83] Matthews; L. Schele
[84] C. H. Rätch.
[85] Rosenzweig, 235; Riva Palacio, 236; Porter, 166.
[86] Riva Palacio, Vol. I.
[87] Porter, 154; Riva Palacio, Vol. I.
[88] Robicsek, 9.
[89] Heimann, 20.
[90] Brooks, 17; Robicsek, 9–11; Las Casas, 212–13.
[91] Robicsek, 11.
[92] B. Díaz de Castillo, 38.
[93] Von Hagen, 242.
[94] Victor W. Von Hagen and Rosenzweig.
[95] Von Hagen, 242–46.
[96] Von Hagen, 241.
[97] Von Hagen, 243.
[98] Riva Palacio, 320.
[99] Robiczek, 12.
[100] Von Hagen, 241.
[101] Robiczek, 41.
[102] Robiczek, 14–22, 27–36, 37–43.
[103] Rosenzweig, 388—89
[104] Riva Palacio, 342; Von Hagen, 241.
[105] Rätch, 170; Von Hagen, 239.
[106] Robicsek, 14, 112; Riva Palacio, 307–09; Recinos' book, 1947.
[107] Robicsek, 112, Riva Palacio, 307–09.
[108] H. Berlin, 1965.
[109] Robicsek, 3, 14–16.
[110] Robicsek, 14–16.
[111] Robicsek, 12–13.
[112] Riva Palacio; Mathews-Schele.
[113] Mark Cusubel.
[114] Robicsek, 14.
[115] R. Bonzel, 1952; J. Tensell, 1970; Starr, 1902, 71.
[116] Seler, 1902, 1:550; Robicsek, 14.
[117] Porter, 171.
[118] Thompson, JES.
[119] Corlett, 184; Nordenskjolde, 184–86; Robicsek, 23.
[120] Spinden, 84; Roys, 1967, 97.
[121] Riva Palacio, 644.
[122] Tezozomoc book.
[123] Adrián Recínos, 221.
[124] Aguírre, 129.
[125] Recínos, 221.
[126] Tubular pipes made of clay, wood, onyx or other stones, or simply reeds.
[127] B. Díaz del Castillo is not correct in his descripition since the word "tobacco" is a Taino word. The Aztecs' name for tobacco was "yetl." Diaz wrote his memoirs in 1585 in order to correct inaccurate accounts of the conquest of México. The word "tobacco" was in common use when he wrote his memoirs.
[128] *Elxochicocozot*, commonly known in México, as "liquidambar," is a resin obtained from a large tree that bears fruit with spikes that have an odd polygonal shape, a black surface and yellow edges. By cutting the bark of this tree, a resin is obtained that when dried and mixed with tobacco sweetens its flavor.The tobacco in use by the Aztecs was very bitter (nicotiana-rustica). Porter, 172.

[129] Prescott, 117–18, 122–27.
[130] Porter, 154.
[131] Porter, 166; Tezozomoc, 211–42; Durán, 161–62, 237, 241, 296–307.
[132] Aguírre, 128–52; Porter, 166.
[133] Porter, 171–72; Duran, 334; Sahagún, 3:71.
[134] Aguírre, 128, 152, 263; H. Osmond, 1955; P. Furst, 1972; Corlette, 169.
[135] Aguírre, 128–29.
[136] Aguírre, 129.
[137] Chagon, 131.
[138] Aguírre, 129.
[139] Aguírre, 120; F. Hernández, 1942.
[140] Heimann, 23.
[141] Heimann, 15.
[142] Adrián Recínos, 221.
[143] Porter, 144.
[144] Torquemada, 1–168.
[145] Sahagún, 11–158.
[146] Tezozomoc, 162, 235–36, 242, 395–410.
[147] Corlette, 109.
[148] Rosenzweig, 16; A. G. Miller, 220.
[149] Sahagún, 189–253.
[150] Sahagún, II, 189, 253.
[151] Rosenzweig, 389.
[152] Rosenzweig, 389; Las Casas, 562–97; Sahagún II: 151, 158.
[153] Corlette, 169; Brooks, 28–29.
[154] Cruz and Badianus manuscript at the Vatican.
[155] Corlette, 172–73.
[156] E. Nordenskjold, 1930; Corlette, 16; Sustell, 154.
[157] Porter, 146.
[158] From Clavijero, cited by Porter, 168.
[159] Porter, 165–66; Aguírre, 128–29.
[160] Noguera, 89–103.
[161] Schondube, personal communication.
[162] Alvin, 213; Moedano, 21–42.
[163] Porter, 203.
[164] Scondube, personal communication.
[165] Porter, 169.
[166] Porter, 169–70.
[167] Porter, 166.
[168] Porter, 170.
[169] Porter, 170.
[170] Archeologist Otto Schondube, Curator of Guadalajara Museum of Anthropology, personal communication.
[171] MacLeod, 574–75; Porter, 155–76; Sir C. Mackenzie, 1957; V. T. Harlow, 1971.
[172] Porter, 55.
[173] Spinden, 57.
[174] Porter, 176.
[175] Oyarzon, 414–37.
[176] Brooks, 32.
[177] Rosenzweig.
[178] Heimann, 8.
[179] Brooks, 15.
[180] Brooks, 5–15; Heimann, 28; MacLeod, 574–75.
[181] Ehwa, 19–24; Heimann, 8–9; Brooks, 18.
[182] "Pigafetta's Narration of Magellan's Voyage," *American Heritage* Vol. XX, No. 6 (1969).
[183] Heimann, 9.
[184] Heimann, 13–14; Spinden, 57.
[185] Chagnon, 37, 131; MacLeod, 574–75.
[186] Chagnon, 23–25.

[187] Chagnon, 23–37.
[188] V. T. Harlow, 99.
[189] Sir C. Mackenzie, 1957.
[190] V. T. Harlow, 1971.
[191] V. T. Harlow, 192.
[192] MacKenzie, 84–85.
[193] MacKenzie, 84–85.
[194] Heimann, 14.
[195] Heimann, 15.
[196] Mortimer, 158; Prescott, 4.
[197] Mortimer, 158; Prescott, 1:140.
[198] Alvin M. Joseph Jr., *The Native Americans.*
[199] Prescott, 1:140.
[200] Prescott, 140.
[201] Collete.
[202] Nordenskjolde, 1930.
[203] Spinden, 57.
[204] Mortimer, 208.
[205] Mortimer, 209.
[206] Corlette, 228.
[207] Mortimer, 212.
[208] Valdéz, 16th century; Mortimer.
[209] Mortimer, 177.
[210] Mortimer, 177–79.
[211] Mortimer, 180.
[212] Mortimer, 182.
[213] MacLeod, 574–75; Porter, 154–55, 174–79; Bowman, 179.
[214] Porter, 154–55.
[215] Porter.
[216] McGuire, 351–645.
[217] Porter, 158.
[218] E. Morison.
[219] E. Morison, *AOS,* 368–74.
[220] Morison, 316.
[221] E. Morison, *Northern Voyage,* 256–57.
[222] Morison, *NV*, 289.
[223] Morison, *NV*, 289.
[224] Morison, *NV*, 277.
[225] Morison, 295.
[226] Morison, *NV*, 298.
[227] Heimann, 16.
[228] Morison, *NV*, 418.
[229] Morison, *NV*, 428; Brooks, 18.
[230] Penn, 9–10; Morison, 418.
[231] Morison, 419.
[232] Riva Palacios, 232–34.
[233] Heimann, 25.
[234] Maxwell, 222–30.
[235] Maxwell, 208.
[236] Maxwell, 228–29.
[237] Lorant, 6.
[238] Lorant, 7.
[239] Lorant, 8.
[240] Lorant, 30–86.
[241] Lorant, 75.
[242] Porter, 182.
[243] Maxwell, 34–30.
[244] Maxwell, 63.
[245] Maxwell, 80–109.
[246] Maxwell, 81.
[247] Lorant, 122.
[248] Morison, 618.
[249] Hakluyt, 1589; Heiman-Laurant, 16.
[250] Sir Richard Greenville, 1585.
[251] Morison, 642–45.
[252] Morison, *AOS,* 647.
[253] Morison, *AOS,* 642; Lorant, 246.
[254] Lorant, 246.
[255] Lorant, 246; Morison, *AOS,* 642.
[256] Lorant, 225; Morison, *AOS,* 642.
[257] Maxwell, 152.
[258] Hakluyt, as reported by Brooks, 49.
[259] Heimann, 135–36; Moya Pons, 38, 46–47, 82.
[260] Moya Pons, 38, 46–47, 82.
[261] National Geographnic, Sept. 1987, 375.

[262] Brooks, 110.
[263] Maxwell, 21.
[264] Maxwell, 124.
[265] Maxwell, 111–12, 114.
[266] Maxwell, 111–12, 114.
[267] Maxwell, 114.
[268] MacCraken, 13–20.
[269] MacCraken, 177.
[270] Maxwell, 142.
[271] Barrett, 117–376.
[272] Epes, 105.
[273] Drysdale, 12–32.
[274] Miller, *Custer's Fall* (London, 1965), 170–71.
[275] Heimann, 127.
[276] Heimann, 28.
[277] Heimann, 30.
[278] Maxwell, 142.
[279] Porter, 152–53.
[280] Heimann, 16; Maxwell, 142; McCracken, 154–77; Ewha, 28.
[281] Heimann, 16.
[282] Barnard, 34.
[283] Barnard, 34–39.
[284] Barnard, 34–39.
[285] Barnard, 34.
[286] MacCraken, 28–35; Barnard, 38–39; Heimann, 121–27.
[287] Heimann, 122.
[288] Porter, 159.
[289] Porter, 160.
[290] Heimann, 123.
[291] Bergon, 334.
[292] Bergon, 326–69.
[293] Dayton Duncan, *Lewis and Clark: The Journey of the Corps of Discovery,* 171.
[294] Heimann, 130.
[295] Porter, 158; Wissler, 221.
[296] Porter, 157–58.
[297] Heimann, 130.
[298] Beechey, 411–12.
[299] E. Morison.
[300] Maxwell, 295.
[301] Fortuín, 132.
[302] Kiernan, 11.
[303] Maxwell, 295.
[304] Heimann, 127.
[305] Heimann, 127.
[306] Heimann, 126.
[307] Franchiére, 107.

Part II

[1] Corti, 58.
[2] Corti, 50.
[3] Rerum Medicaron, Novae, 16 volumes on his book Cuarto Libros published in 1615.
[4] Corti, 1559.
[5] Brooks, 46.
[6] Corti, 65.
[7] Corti, 57.
[8] Corti, 60.
[9] Corti, 59.
[10] Corti, 62–63.
[11] Corti, 55.
[12] Brongers, 6–25.
[13] Corti, 55.
[14] Corti, 56–57.
[15] Brooks, 125; Corti, 58.
[16] Brooks, 34, 75–76, 274.
[17] Brooks, 34.
[18] Brooks, 209, 76–77.
[19] Brooks, 282.
[20] Brooks, 210.
[21] Brooks, 61–62.
[22] Brooks, 62.

[23] Corti, 68.
[24] MacKenzie, 84–85.
[25] Corti, 69.
[26] Corti, 69–70.
[27] Corti, 69.
[28] Corti, 68.
[29] Brooks, 62.
[30] Brooks, 63.
[31] Brooks, 67–68, 50–51.
[32] Brooks, 46–48.
[33] Edmund Spenser, Book 3, Cantos. 3.
[34] Brooks, 72–73, 282.
[35] *See* Basil Davidson, *The African Slave Trade* (Little Brown).
[36] Brooks, 9, 48, 58, 113.
[37] Brooks, 93–99; *Basil Book of Black Slavery.*
[38] Brooks, 49.
[39] Brooks, 53.
[40] Brooks, 7, 53.
[41] Brooks, 57.
[42] Brooks, 53.
[43] Brooks, 54–55, 156.
[44] Brooks, 53–55, 224–25, 237, 284.
[45] Brooks, 54–57, 87, 96, 99, 112, 151, 154.
[46] Brooks, 163–64.
[47] Brooks, 93.
[48] Brooks, 166–67.
[49] Heimann, 67–68.
[50] Brooks, 89.
[51] Brongers, 24–25.
[52] Corti, 98–99.
[53] Brongers, 25.
[54] "Tobacco is lauded to the skies and recommended as a remedy for all the ills to which the flesh is heir." (Johananes Neander of Bremen, *Tabacologies,* (Leyden, 1622). Corti, 104–05.
[55] Brook, 44.
[56] Brooks, 43–44.
[57] Brooks, 78, 282.
[58] Brooks, 47, 75, 78, 130, 205–07.
[59] Laufer, 63; H. Phillips, 1882.
[60] Laufer, 160.
[61] Corti, 139–40.
[62] Corti, 145–46.
[63] Brooks, 3, 209.
[64] Corti, 48.
[65] Brooks.
[66] Brooks, 34.
[67] Brooks, 114–15.
[68] Brooks, 210.
[69] Corti, 144.
[70] Brooks, 144, 09.
[71] Brooks, 69.
[72] Corti, 94–95.
[73] Corti, 166.
[74] Brooks, 3, 203.
[75] Brooks, 3–4.
[76] Brooks, 75.
[77] Brooks, 79.
[78] Landa.
[79] Brooks, 79.
[80] Kiernan, 102.
[81] Corti, 129–30.
[82] Corti, 129–30.
[83] Corti, 113.
[84] Brooks, 141.
[85] Corti, 135.
[86] Brooks, 77–78, 151.
[87] Mark, VII, 15.
[88] Corti, 142–43.
[89] Brooks, 152–53.
[90] Laufer, 59–60.
[91] Brooks, 255.
[92] Brooks, 75–76, 230.
[93] Brooks, 273–74.
[94] Brooks, 258.
[95] Kiernan, 30; A. G. Bronger, 31–48.
[96] Brooks, 9, 70–71.

[97] A. G. Bronger, 71–78.
[98] Ferrier, 241.
[99] The Two Commissars, Bombay.
[100] Brooks, 8, 209.
[101] J. P. Ferrier.
[102] McClellan.
[103] Brooks, 128.
[104] Corti, 198.
[105] Bronger, 158.
[106] Corti, 202.
[107] Landon 441.
[108] Brooks, 129.
[109] Brooks, 302.
[110] Books, 138.
[111] Brooks, 141.
[112] Corti, 211.
[113] Liselotte to the Raugrafin Louise, 1713.
[114] E. Anderson, 3rd edition (London: 1985).
[115] Kiernan, 108.
[116] Grogan.
[117] Brooks, 213, 220, 235, 244, 266, 277.
[118] Brooks, 217, 301–02.
[119] Brooks, 277.
[120] Brooks, 109.
[121] *A Weekly Trader* review on tobacco in 1946 (Brooks, 332).
[122] Brooks, 234.
[123] Brooks, 221, 245.
[124] Brooks, 274.
[125] Conrad, 35–40.
[126] Conrad, 40–41.
[127] Conrad, 43.
[128] Conrad, 44.
[129] Corti, 255, 257.
[130] Corti, 214–19.
[131] *L'etre et Le Neant.*
[132] Brooks, 85.
[133] Brooks, 145.
[134] Corti, 253.
[135] Corti, 252.
[136] Brooks, 234.
[137] Hoffman, 125–27.
[138] Ravenholt, 21.
[139] Ravenholt, 220.
[140] Hamilton, This Smoking World, 227; Corti, 259–64.
[141] C. D. Memosin.
[142] Ravenholt, 63.
[143] Oscar Auerbach, *New England Journal of Medicine* Vol. 256 (January 17, 1957), 97–104.
[144] In 1963 I had the privilege and pleasure of taking a trip to Jamaica with Dr. Oscar Auerbach and Dr. Ernst Wynder, where both of these physicians presented a paper on the very important research that they had done in relation to tobacco as a causative factor in cancer of the lung. Dr. Auerbach collected old American quarters (coins) as a hobby. I was baffled for a long time, and even am now, why such a scientist with such deep knowledge would direct his attention to something so trivial. Dr. Auerbach insisted on showing me a collection of coins that he had recently acquired and that he had brought in his pocket to the conference. I became distracted and started smelling the wonderful scent of the blue mountains where we were invited to a dinner. It is strange and unexplainable how the mind works sometimes. On the other hand, Dr. Wynder, who was a very serious man, had such a good time enjoying a dance with the beautiful women of Jamaica.

BIBLIOGRAPHY

PART I

A

ACOSTA J.	*1936.*	Concilium Limensis 1591 in G. Arents Jr. G. A Tobacco History 1937.
AGUÍRRE B.G.	*1963.*	Medicina y Magia Instituto Nacional Indigenista México D.F.
ALVIN M Jr. J.	*1966.*	The Indian Heritage of America Alfred A. Knoff Inc. NY. NY
ALVIN M Jr. J	*1961.*	The American Heritage. Book of Indians Simon and Schuster New York, NY
AMERICAN HERITAGE	*1969.*	Vol XX No 6. a) The End of the Iroquoi b) Pigafetta's narration of Magellan's voyage
AMERICAN HERITAGE. *d) from Bandelier, Fanny's Translation.	*1963.*	HISTORY OF THE UNITED STATES a) French Trade with America under Louis XIV. P.14 b) Dutch settlement in New York. P. 48 c) Queen Elizabeth and James I XVI-XVIIC. P. 49 *d) The Ordeal of Cabeza de Vaca 1528. pgs. 83–88 e) The five seals of the colonizers. Front page
ARANGO F.	*1991.*	Boletín. Museo del Hombre Dominicano Año XVIII. Num 24. Sto Domingo. Dom. Rep.
ARDEN H.	*1987.*	National Geographic Magazine On Iroquoi Indians—Confederacy. 'The fire that never dies' Sept 1987 (pg. 375)

ARENTS Jr. G. *1937.* Tobacco Its History
The Books and Manuscripts and Engravings
In the Library of George Arents Jr.
The Rosenback Co.
New York. MCMXXXVIII. Vol I (1507–1615)
(Data obtained at the New York Public Library 1996)

ARVIGO R & BALICK M 1993 Reinforest Medicines. One Hundred Healing Herbs of Belize (Maya)
Lotus Express. Twin Lakes WI

B

BACHILLER Y MORALES A 1883 Cuba Primitiva, Origen Lenguas, Tradiciones y Historia De Los Indios De Las Antillas Mayores y Las Lucayas
Habana 1883

BANCROFT 1875 The Indian of México also snuffed and chewed Tobacco.
Bancroft 1875 (pg. 287) Stuhl, 1932 p. 105

BAYER H. 1936 Botany of the Maya & Miscellaneous papers.
Carnegie Institution of Washington
Publication No. 461. Washington D.C.

BAYER H. 1976 Ritual Cloth and Palenque Kings (De La Fuente 1965)
Proceedings segunda mesa redonda de Palenque.
Robert Louis. Stevenson School. Ed. M. Green Robertson. Pebble Beach California.

BENZONI G. *1568.* Historia Del Nuovo Mundo. Venice BK4,pp 116

BERLIN H. *1965.* The Inscription of The Temple of The Cross at Palenque. American Antiquity. 30: 33–42

BESSON M. *1929.* The Scourge of the Indies.
Buccaneers, Corsairs, Filibusters.
Random House NY

BLOM F. *1955.* La Selva Lacandona. México City.

BOOK OF Chilam Balam of Kanda. MS copy in Latin American Library. Tulane University New Orleans

BONZEL R. *1952.* Chichicastenango. Seattle University (Guatemala). Washington Press.

BRASSEUR de BOURBOURG *1861.* Popol Vuh. Le Livre Sacre et Les Myths De L'Antiqua Americaine Avec Le Heroique De Quichés. Paris. A. Bertrand

BRENER LEAH	1963	An Artist Grows Up in México. The Beechurst Dress Inc. 1953 (Pauleo)
BRONGERS G.	1964.	Nicotiana Tabacum. GTSH Museum Gronigen Netherlands.
BROOKS J.	*1952.*	The Mighty Leaf (Book) Little Brown and Co. Boston, Mass.

C

CÁRDENAS J.	*1913.*	Primera Parte de los problemas y secretos maravillosos de las Indias México D.F. Imp. Del Museo Nacional (from a 16th C. manuscript. Ref by Porter) 1913
CAPRETTI ELENA	2003	Bruneleschi Gunti p. 28 (Book)
CHAGNON A N.	*1968.*	Yąnomamö (Yanomami). The Fierce People Stanford University Holt Reinhart el al.
CHAVÉRO A.	*1930.*	Chilam Balam de Chumayel. Translated by Antonio Mediz Bolio . San Jose Costa Rica Imp. Y Librería Lehmann.
CHEVALIER	1863	Le Mexique ancient et Moderne (Pom 1863)
CHILAM BALAM OF KAUA		Copy in the Latin American Library. Tulane University. New Orleans. Ms. Gates. Reproduction in Peabody Museum. Harvard University. Cambridge, Mass.
COBO BERNABÉ	1890– 1895	Historia del Nuevo Mundo. 4 vols. Seville 1890–1895.
CODICE BORGIAN	1904	Eduardo Seler. Ed. Barlin
CODEX TROCORTESIANO (Madrid Codex)	1930	Códices Mayas por J. Antonio Villacorta Guatemala CA. Tipografía Nacional Códice Troano taken from Ms. De Brasseur de Bourbourg. Manuscripto Troano de París 1869–1870
COHEN B.	*1992.*	What Columbus "saw" in 1492 Scientific American 1992
CORLETT W.T.	*1935.*	The Medicine Man of the American Indian Charles C. Thomas
CORTI C.	1929	Maximillian and Charlotte of México.
CORTI C.	*1932.*	A History of Smoking translated from French to English by Paul England Harcourt Brace Co. 1932
CRÓNICAS DE AMERICA		Cantos y Crónicas del México Antiguo.

D

DAVIDSON BASIL	1980.	The African Slave Trade Little Brown Co.
De La Fuente B.	1965 1976	La escultura de Palenque. Universidad Autonoma de México. México City Plate 50
DÍAZ del CASTILLO B.	1973.	The Conquest of New Spain. Middlesex
DIXON R.B.	1921	Words for tobacco in American Indian languages American Antropologist N.S. 23:19–49, 1921
DIXON R.B.	1933	American Antropologist vol. 35. Tobacco chewing In the North West Coast. Lancaster pg. 1933
DRYSDALE V.L.	*1953.*	The Gift of the Sacred Pipe. The Sacred Calumet Pipe and the Great Spirit Joseph Epes Brown. University of Oklahoma Press. Norman.
DURÁN F. DIEGO	*1951.*	Historia De Las Indias De La Nueva España y Islas De Tierra Firme. Ed. J.F. Ramírez. 2 vol. Atlas. México City Editorial Nacional
DYSON J.	*1991.*	For Gold God and Glory Book on Colombus. Simon and Schuster Madison Press Books. New York

E

EASTWOOD E.	1938	The Tobacco Collected by Archival Menzies on the North West Coast of North America. (Leaflets of Western Botany, no. 6) San Francisco 1938.
EHWA Jr. C.	*1974.*	The Book of Pipes and Tobacco A Ridge Press Book Random House New York
ENCYCL. BRIT.	*1982.*	Macropedia. Vol. 18 15th Edition
ENCYCL. BRIT.	*1976.*	The Annals of America Vol. I—pg. 11 Discovery of the New World
ENCYCL. BRIT. (T)	*1982.*	Tobacco. pg. 464 . pgs. 111–118
ERNST A.	*1889.*	On the Ethymology of Tobacco American Anthropologist Antigua Serie, 2: 133–141

ESTIENNE C.H. & JEAN LIÉBAULT	1567	L'Agriculture Et Maison Rustique. Paris 1567
EWERS J.C.	*1957.*	Black Foot Indian Pipes and Pipe Making. Smithsonian Instiution Bulletin 186. Washington D.C.

F

FERGUSON W.	*1977.*	Maya Ruins of México in color University of Oklahoma Press
FERNÁN I.F.	*1976.*	Tobacco y Sociedad Centro de Investigación y Acción Social Dominican Republic. (Chapter I pgs. 29–41)
FORTUINE R.	*1991.*	Enjoying the Weed Arctic Style (1826–27) Alaska Medicine. Glimpses of Alaskan Medical History. 33(3).132 Jul-Sept.
FRANCHEERE G.	*1969.*	Journal of a Voyage on the North West Coast of North America 1820. Toronto 1969. On the Chinook Indians
FUNDABURK E.	*1957.*	Sun Circles and Human Hands The Southeastern Indians Art and Industry Elf Luverne. Alabama.
FURST PETER T.	*1972.*	Flesh of the Gods. The ritual use of Hallucinogens (book).
FURST PETER T.	*1974.*	Archeological Evidence of Snuffing in Prehispanic México. Botanical museum Leafets. Harvard University vol. 24, No. 1 pp. 1–28. Cambridge, Mass.
FUSON R.	*1987.*	The Log of Christopher Columbus (Translated from the original log by F.R.) International Marine Publishing Co. Camden Maine.

G

GADE J.A.	1942	Chrisitian IV of Denmark and Norway (XVIIC). London 1927
GOODSPEED T.H.	1941	Plant hunters in the Andes. NY 1941 Chromosome number and morphology in Nicotiana VII Karo-types of 55 species in relation to a Taxonomic revision of the genus: University of California. Botanical studies. Vol. 18 Berkeley California 1945.

GUMÍLLA 1741 El Orinoco Ilustrado—Ed. P.T. Fiest New York 1972.

H

HALL R.F. 1973 Reference on different kinds of tobacco.

HALL R.F. 1975 On Kinnickinniyc Mound Burial . . . MS. University of Illinois at Chicago Circle Chicago.

HALL R.F. 1975. Ideology and Adaptation in North America MS University of Illinois, Chicago Circle Chicago

HARLOW V.T. *1971.* Raleigh's Last Voyage. Da Capo Press NY Argonaut Press #12

HARRINGTON J.P. 1932 Tobacco among the Kuruk Indians of Calfornia. (Bureau of American Ethnology. Smithsonian Institution Bulletin No. 94) Washington 1932.

HEIMANN R.K. *1960.* Tobacco and Americans (Book) McGraw Hill Co. Inc. New York

HEIZER R.F. 1940 Botanical Identification of North West Coast Tobacco (American Anthropolgist vol 42) Lancaster pg. 1640

HERNÁNDEZ F. *1945.* Historia de las plantas de la nueva España. (Antigüedades de la Nueva España. Traduccion del latín y notas Por Joaquín. García Pimentél, México 1945) Méxican Edition 1945

HERODOTUS *1855.* Herodoti Historiae. New York: Henry Clay

HOLTON P 1977 The Works of Jacques Le Moyne By the Trustees of the British Museum 1977

HOMBERGERE 1995 The Penguin Historical Atlas of North America. Viking 1995.

HUMBOLDT A. & BONBLAND G. 1818 Nova Genera of species plantarum, Paris, 1818

I

IRVING W. *1896.* Columbus. His life and voyages G.P. Putnam's Sons New York

ISRAEL J.I.	*1989.*	Dutch Primacy in World Trade (1585–1740) Oxford 1989. (Ref. In Kiernan's Bibliography)

K

KIERNAN V.G.	*1991.*	TOBACCO. A History Hutchinson Radius. An Imprint of Random Century Ltd. London—Australia.

L

LAFARGE B		Quoting the Popol Vuh 123–24
LANDA D Fr.	1928	Relación de Las Cosas de Yucatán. México City
(16th century)	1941	(Translation by Tozzer. Peabody Museum of Archeology and Ethnology. Harvard University. Vol. 18. Cambridge 1941
LAS CASAS F.B.	*1985.*	Historia de Las Indias vol. I Adiciones del Continente by Roberto Cassa
LE MOYNE JACQUES *De MORGUES*	*1977.*	The Work of Jaques Le Moyne The Trustees of the British Museum British Museum Publications Limited 1977.
LEÓN N.		Los Tarascos. México DF. Imprenta Del Museo Nacional (Ref. By Porter)
LISTON R.	1924	Use of tobacco among North American Indians Field Museum of Natural History Anthropology Leaflet No. 15. Chicago.
LÓPEZ-AUSTIN A		Florentin Codex. Sahagún.
LORANT S.	*1946.*	Los Tarascos. Imprenta del Museo Nacional México D.F.
LORANT S.	*1946.*	The New World. The First Pictures of America made by Jolin White and Jacques LeMoyne. Engraves by Theodore de Bray. Duell Sloane and Pearce N.Y.
LOTHROOP S.K.	1924	Tulúm. An Archeological Study of the East Coast of Yucatán. Carnegie Institution in Washington D.C.

LOVÉN SVEN 1935 Origin of the Taino Culture. West Indies. Göteborg 1935.

LOWIE R.H. 1919 The Tobacco Society of the Crow Indians. American Museum of Natural History. Anthropological papers. Vol. 21. New York

LOWIE R.H. 1956 The Crow Indians. New York

LUMHOLTZ K. 1898 The Huichol Indians of México (American Museum of Natural History. Bulletin vol. 10) New York, 1898

M

MACKENZIE SIR C. 1957. Sublime Tobacco On Raleighs account of his voyage to the Orinoco River. London

MACLEOD W.C. 1930. The Chewing of Tobacco on Southeastern North America American Anthropologist 32: 574—575 1930–

MASON J. ALDEN 1924 Use of tobacco in México and South America. Field Museum Leaflet 16. Chicago.

MATTHEWS P. 1973 Lords of Palenque, The glyphic
SCHELE L. 1974 EVIDENCE. Proceedings Primera Mesa
1976 Redonla de Palenque. Vol. I. Robert Louis Stevenson School. Ed. M. Green Robertson Pebble Beach, California

MAXWELL J. 1978. America's Fascinating Indian Heritage Reader's Digest Association Inc. Pleasantville, NY

MCQUIRE J.D. 1896– Pipe and smoking of the American Aborigines National Museum Annual Report. Publication No 1899 Washington D.C. 1896–1897

MILLER, A.G. 1972. The Mural Paintings of Teotihuacán. Dumborton Oaks. Trustee of Howard University.

MILLER, M. 2004 Courtly Art of the Ancient Maya
SIMON MARTIN Fine Arts Museum of San Francisco and California Palace of the legion of honor

KATHLEEN BERSIN		Thames and Hudson
MOEDANO H.	*1941.*	Estudio Preliminar de la Ceramica de TzinTzunTzan 139–1940 Revista Mexicana de Estudios Antropológicos México DF.
MONARDES N.	1565	Historia Medicinal de Las Cosas Que Sirven Al Uso de La Medicina. Seville, Spain (Ref.Porter 141)
MORBÁN L.F.	*1996.*	On Customs of the Taino Indians. Buletín del Hombre Dominicano Personal Comunicatoon.
MORISON S.E.	*1942.*	Admiral of the Ocean Sea (AOS) Little Brown and Co. (book).
MORISON S.E.	*1971.*	The European Discovery of America The Northern Voyages (N.V.) New York Oxford. University Press (book).
MORLEY G.S.	*1983.*	(Revised by Robert Sharer) 4th Ed. Stanford University Press Palo Alto, Ca.
MORTIMER W.G.	*1974.*	History of Coca Fitz Hugh Ludlow Memorial Library Edition San Francisco, Ca
MORTON J.	*1968.*	The Calabash (Crescentia cujete) in Folk Medicine. Economic Botany 22:273–80
MOYA PONS F	1977	Manual de Historia Dominicana. Academic Dominicana de La Historia. Santo Domingo, Dominican Republic
MUÑOZ C.	1892	Historia de Tlaxcala México D.F. Alfredo Chávez.

N

NATIONAL GEOGRAPHIC	2002	Diggins in Ft. James Town (June issue) Unsettling Discoveries at Jamestown By William Kelso & Karen E. Lang. Photograph by Ira Block pg. 74–81
NAVARRETE D.F.	*1890.*	Die Reisen des Cristoph Columbus 1492—1504 Leipzig.

NOGUERA EDUARDO	1931	Exploración Arqueológica en las regions de Zamora y Pátzcuaro Edo. De Michuacán. Anales Museo Nacional de Arqueología y Etnología. Ep 47:89–103 México DF 1931
NORDENSKJOLD E	*1919.*	The use of enema tubes and enema Syringes among the Indians Ethnographiska Studier, no. 8, pp. 184–86. Göteborg 1919 An ethnographic analysis of the material culture of two Indian tribes in the Gran Chaco Göteborg 1919.

O

OBREGÓN M.	*1987.*	The Barcelona Letter of 1493 The Landfall Controversy and the Indian Guides MacMillian Publishing Co. New York 1987.
OBREGÓN M.		The Columbus Logs.
OSMOND H	*1955.*	Ololiuhqui. The Ancient Aztec Narcotic. Journal of Mental Science 101: 526–37
OVIEDO F.	*1535–* 1851	Historia General de las Indias Islas y Tierra Firme del Mar Oceano Madrid. Publication de la Real Academia de la Historia Madrid Spain. From Oviedos' History 1535.

P

PANÉ RAMÓN	1571	Scrittura Di Fra Roman Delle Antichità de GL Indiani (In: Columbo Fernando. Historie Del S.D. Fernando Colombo, Venice, 1571)
PENA R. DE L'OBELL M	1570–1571	Stirpium adversaria nova. London 1570–1571.
PENN W.A.	*1901.*	The Soberane Herb London Gram Richards New York : E.P. Dutton and Co. 1901
POMAR J.B.	*1903.*	Relación de las ceremonias de la Provincia de Michuacán. Virrey de la Nueva España. Sr. Don Antonio Mendoza 1903

PORTER M.N.	*1948.*	Pipas Precortesianas. Acta Antropologica 111.2 México DF. June 1948
PRESCOTT W.H.	*1847.*	History of the Conquest of Perú—Vol II First Edition Harper and Brothers 82 Cliff St. New York
PRESCOTT W.H.	*1843.*	Conquest of México. Vol II. (First Edition.) Harper and Brothers 82 Cliff St. New York

R

RÄTCH CH.	*1957.*	Plants of Love. Ten Speed. Berkeley Calif. (Balche the sweet medicine from Heaven.)
RECÍNOS A.	*1947.*	Popol Vuh. Antiguas historias de Quiché. México City.
RECÍNOS A.	*1947.*	Popol Vuh. The Sacred Book of the Ancient Quiché Maya. Norman. University of Oklahoma Press.
RECÍNOS, GOETZ AND MORLEY	*1947.*	Popol Vuh. The Sacred Book of the Ancient Quiché Maya. Norman. University Of Oklahoma Press.
REKO BLAS	*1903.*	Relación de Mechuacán. Ceremonias Y Ritos y población y gobernación de la Chumayel. Norman. University of Oklahoma Press.
RIVA-PALACIO V.	*1886.*	México a travez de los Siglos TOMO I (Vol I) First Edition
ROBELA C.	1911	Diccionario de Mitología náhuatl. México 1911.
ROBICSEK F.	*1978.*	The Smoking Gods. Tobacco in Maya Art. University of Oklahoma Press. Norman.
ROSEN ERICGBVON	1924	Ethnological research work during the Swedish—Chaco-Cordillera Expedition 1901–1902 Stockholm 1924.
ROSENZWEIG- DÍAZ A.	*1957*	Mexicanidad de México Vol. I. Editorial The Dolphin Book Co. Ltd. Oxford.

ROTH H.L. 1887 The Aborigines of Hispaniola (Royal Anthropoligical Institution Journal, Vol. 16. London 1887)

ROYS R.L. 1931. The Ethnology of the Maya. Tulane University, 1931.

ROYS R.L. 1967. The Book of Chilam Balam of Chumayel. University of Oklahoma Press (1933). Norman.

RUÍZ DE MONTOYA R. 1640 Arte y Bocavulario De La Lengua Guaraní. Madrid, 1640.

RUZ L. 1950 Exploraciones en Palenque. Temple of the Cross in Palenque. Plate 50

1973 Temple of the Inscriptions.

S

SACA JOSÉ A. 1938 Historia de la Esclavitud De La Raza Africana En El Nuevo Mundo. 4 vol. La Habana, 1938

SAHAGÚN FRAY B. 1982. Historia General de las cosas de la Nueva España.- By Alfredo Lopez Austin y Josefina García Quintana Fomento cultural Banamex. 1982.

SAHAGÚN FRAY B. 1938 History of the New Spain.

SELER E. 1902–1903, 1937 Codex Vaticanus. No 3773—an old Mexican poetrial manuscript in The Vatican Library. Translated.. H.H. Keane Bartuor

SETCHELL W.A. 1921 Aboriginal Tobacco American Anthropologist 23:397–414

SILVER, D. 1966. Zinacantán Shamanism, PhD Dissertation. Howard University.

SMITH J. 1627. The General History of Virginia. London. (Encyclopedia Brit).

SOUSTELLE J. 1962. Daily Life of the Aztecs. The MacMillan Co. 1962.

SPINDEN H. 1950. Tobacco is American The New York Public Library.

STANLEY P.C. 1930 Flora of Yucatán. Field Museum of Natural History Botanical Series vol. 3 no. 3 Chicago 1930.

STARR F. 1900–1901 Notes on the Ethnography of Southern México. Proceedings of Davenport, Iowa. The Davenport Academy of Natural Sciences. Vol. II.

STIRLING M. *1965.* Indians of the Americas
National Geographical Society
A Color Illustrated Record
by The National Geographic Society

SWITZER H.A. *1969.* Tobacco, Pipes and cigarettes of the Prehistoric Southwest, El Paso, Texas

T

TENZELL J. *1970.* Shamanism and concepts of disease in a Mayan Indian community. Psychiatry 33: 322–80

TEZOZOMOC D.H.A. *1878.* Códice Rodríguez
Tezozomoc version of Aztec History from just after the conquest of México (XVIC)
Cronica Mexicana First Ed.
Orozco Y Berra; Editors. México DF.

THEVET A. *1558.* Le Singularités de la France Antarctique, autrement nommée Amerique, Paris.

THOMAS HUGH 2003 El Imperio Español (Book)
Planeta Historia y Sociedad Barcelona Spain. 2003

THOMAS JOYCE 1927 Maya and The Mexican Art Lords.

THOMPSON J.E. *1939.* The Moon Goddess.

1946 Some Uses of Tobacco among the Maya
Carnegie Institution of Washington.
Notes on Middle American Archeology and Ethnology, Washington, D.C.

THOMPSON J.E. 1970 Maya History and Religion. Norman.
University of Oklahoma Press.

THOMPSON J.E. 1974 The Influence of Psychotropic Flora and Fauna on Maya Religion.
Current Anthropology 15:160, 88–89.

TOZZER A.M. *1966.* Landas Relación de las Cosas de Yucatán.
In Peabody Museum. Harvard Review
Vol. 18 pg. 106. Cambridge, Mass.

V

VEGA B. *1981.* Tabaco y Historia. Estudios Dominicanos.
Vol X. Num 57 Nov / Dec
Estudios Dom. Santiago de los Caballeros.

VEGA B.	1982	Museo del Hombre Dominicano Buletín 17 Año X. Num 17–1982 Sto Dom. Rep. Dom. About few pipes found in Dom. Rep.
VEGA GARCILASO DE LA	1609	Primera Parte De Los Comentarios Reales. Lisbon, 1609.
VOGEL V.	1969	American Indian Medicine. Norman. University of Oklahoma Press.
VON HAGHEN V.	*1961.*	The Ancient Sun Kingdoms of the Americas. Aztec Maya Inca The World Publishing Co. Cleveland, N.Y.

W

WASSEN S.H.	*1964.*	The Use of Some Specific Kinds of South American Snuff and Related paraphernalia. Göteborg.
WASSON R.G.	1963.	Notes on the present status of Ololiuhqui and other hallucinogens of México. Harvard University. Botanical Museum Leaflets 20:161–93.
WASSON R.G.	1966.	Ololiuhqui and the Other Hallucinogens of México. In Suma Antropologica en Homenaje A Robert Weitlander pp. 329–48. México City: Instituto Nacional de Antropologia e Historia.
WEISS G.	*1973.*	Shamanism and priesthood in the light of the Campa Ayahuasca ceremony. In Michael J. Harner, ed. Hallucinogens and Shamanism. New York. Oxford University Press.
WEST G.A.	*1970.*	Tobacco Pipes and smoking customs of The American Indians Westport, Conn.
WILBERT J.	*1972.*	Tobacco and Shamanistic ecstasy among The Warao Indians of Venezuela. In P.I. Furst Book. Flesh of the Gods. Praeger, NY
WISDOM C.	1940.	The Chorti Indians of Guatemala (The University of Chicago publications in Anthropology). Low Lacadon Plantation. (Ethnology series) Chicago 1940

WOODGER ELIN	2004	Encyclopedia of the Lewis and Clark Expedition (pp. 342–343) Checkmark Books.
WOODS C.	1968	Medicine and Culture in San Lucas Tolimán. A Highl and Guatemalan community. PhD dissertation. Stanford University.

Z

SVI DOR-NER	*1991*	Columbus and the Age of Discovery. William Morrow and Co. 1991.

BIBLIOGRAPHY

PART II

BIBLIOGRAPHY: PART II

A

ABBE R.	1915	Cancer of Mouth. Case against tobacco. MJ. 102:1, 1915 N.Y.
ARENTS Jr. GEORGE (Collection)	*1937*	Tobacco—Its History . Illustrated by the Books and Manuscripts and Engravings in the library of George Arents Jr. New York . The Rosenthal Co. MCMXXXVII. *1937*
AUERBACH O.	*1961.*	Changes in the bronchial epithelium in relation to cigarette smoking and in relation to lung cancer. N. Engl J. Med. 1961; 265: 253–67

B

BARNABY CONRAD III	*1996*	The Cigar. Chronicle Books, San Francisco
BESSON MAURICE.	*1929*	The Scourge of The Indians Random House 1929. The broadening of the coast. Filibusters or pirates attacking boarding ships
BILLROTH T.	*1878.*	Lectures on Surgical Pathology and Therapeutics Vol II. Page 471. The New Sydenham Soc. MDCCCLXXVIII

BOLLARD	*1888*	The smoking Craze. An indictment with reason and appeal to christians 3rd Edition (London Partridge) Kerrn (1888) Inebriety. London H.K. Lewis
BRESLOW L.	1914	Occupations and cigarette smoking as factors in lung cancer. AKJ, Public Health 14:171, 1914
BROADEN A.C.	*1920.*	Squamous-cell Epithelium of the Lip. A study of five hundred and thirty-seven cases. JAMA 74 (10) 656–664, March 6, 1920
BRONGERS A. GEORGE.		Nicotiana Tabacum. Theodorus Niemeyer N.V. 1964. Groningen, Netherlands.
BROOKS, J.E.	*1952*	The Mighty Leaf, Boston, Little Brown *1952*

C

CASASOLA, AGUSTÍN V.	*2003*	Essay by Pete Hamill, Edited by Pablo Ortíz Monasterio. México The Revolution & Beyond. Conaculta INAH
CLEMENS J.	1954	Bronchial carcinoma. Danish Medical Bulletin, 1:37, 1959.
CORTI, COUNT	*1932*	A History of Smoking Translated by Paul England. Harcourt, Brace & Co. *1932.*

D

DAFF-DOLL M.E.	1951	Cancer of the Lung in relation to tobacco. British J. Cancer 5:1951
DOLL R.	*1950.*	Smoking and carcinoma of the lung British Medical Journal 1950.
DOLL R.	1953	Bronchial carcinoma. Evidence and etiology. Bristish M.J. 2:521, 585, 1953
DOLL R.	1953	Mortality from Lung Cancer Among Non-smokers. Bristish M.J. Cancer 7:303, 1953
DOLL R.	1954	The Mortality of Doctors in relation to their Smoking Habits. British M.J. 1:1951, 1954

E

EHWA, CARL Jr.	*1974*	The Book of Pipes and Tobacco, A Ridge Press Book, Random House, N.Y. *1974*

ESSENBERG J.M. 1952 Cigarette Smoke and The Incidence of Pulmonary Neoplasms of the Lung in The Albino Mouse. Science, 116:561, 1952.

F

FLORY C.M. 1941 Production tumors by tobacco tars. Cancer Research 1:262, 1941

G

GRAHAM, E.A. *1933.* "Successful removal of an entire lung for carcinoma of the bronchus" JAMA 101, 1371, 1933.

H

HAMMOND E.C. *1966.* Smoking in relation to the death rate of one million men and women. National Cancer Institute Monographs 1966. 19 127–204 Jan.

HAMMOND E.C. *1954.* The Relationship Between Human Smoking Habits and Death Rates JAMA 155 1316 1954.

HEIMANN K. ROBERT 1960 Tobacco and Americans. McGraw Hill Book Co. Inc.—*1960.*

HOFFMAN F. *1746.* A dissertation on Endemial Diseases (original book) Treatise on the Diseases of tradesman London. Printed by Thomas Osborne 1746

HULTON PAUL *1977* The Works of Jacques LeMoyne. By The Trustees of the British Museum *1977*

HUPER W.C. 1947 Occupational Tumors and Allied Diseases. Thomas, Springfield, Illinois 1942.

K

KIERNAN V.G. *1991* Tobacco A History. Hutchinson Radius. Random Centruy LTD London—Australia *1991.*

L

LAUFER BERTOLD 1924 Introduction of Tobacco to Europe Field Museum of Natural History Chicago 1924.

M

MUELLER F.H. *1939.* Tobak Missbrauch und Lungen carcinoma Krebsporsch; 49:57, 1939

N

NEW YORK TIMES *1993.* 'Sin Taxes'. President Clinton's Taxes. Feb. 29, 1993

O

OCHSNER A. *1941.* Carcinoma of the lung Ann of Surg 1941. 42–209

OSLER W. *1892.* The Principles and Practice of Medicine New York. D. Appleton & Co. 1892

P

PARK EDWARD *1983.* Treasures of the Smithsonian Smithsonian Books Harry H.N. Gram Inc. Washington, D.C.

PEARL R. *1938.* "Tobacco Smoking and Longevity Science 87: 216–217.

POTT. P. *1775.* "Cancer Scroti" "In Chirurgical Works of Of Percival Pott". London J.J. Johnson.

R

RAVENHOLT R. *1963.* Epidemiology and Treatment of lung cancer Diseases of the chest. Vol 2. Aug 1963

S

SPINDEN HERBERT J. *1950* Tobacco is American. The New York Public Library *1950*

STANLEY JOHN 1994 Classic Music. A Readers Digest Book (on Paganini)

W

WYNDER L.E. *1950.* Tobacco smoking as a possible Etiologic Factor in Bronchogenic Carcinoma. A Study of 684 proved cases
JAMA May 27, 1950.

Y

YACH D. *1992.* The Columbus Quincentenary 500 Years of Expanding Tobacco Addiction.
South African Medical Journal 1992. 82(4): 224–5